D1204070

GLOBAL FEVER

GLOBAL

How to Treat Climate Change

FEVER

William H. Calvin

THE UNIVERSITY OF CHICAGO PRESS

CHICAGO AND LONDON

William H. Calvin is a professor at the University of Washington
School of Medicine in Seattle. His books can be found at
WilliamCalvin.com and his university web site is at
faculty.washington.edu/wcalvin.

Corrections to this book, as well as downloadable illustrations, can be
found at *Global-Fever.org.*

The University of Chicago Press, Chicago 60637
The University of Chicago Press, Ltd., London

17 16 15 14 13 12 11 10 09 08   1 2 3 4 5

Library of Congress Cataloging-in-Publication Data

Calvin, William H., 1939–
  Global fever : how to treat climate change / William H. Calvin.
     p. cm.
  Includes bibliographical references and index.
    ISBN-13: 978-0-226-09204-1 (cloth : alk. paper)
    ISBN-10: 0-226-09204-6 (cloth : alk. paper) 1. Climatic changes. I.
    Title.
  QC981.8 .C5C342 2008
  363.738'74—dc22
                          2007031376

The outlook is for major complications, such as droughts that just won't quit. Tipping points lead to demolition derbies. The Amazon burns. Major cities drown. Deserts expand. Oceans acidify. Dwindling resources trigger genocidal wars with neighbors (think Darfur). Extreme weather keeps trashing the place.

Absent effective treatment, much of that will be on tap for later this century even if we avoid the most serious problem: sudden flips in climate. The cockroaches and mosquitoes will like our global fever; most of us will not.

What's the treatment? The obvious way to treat the fever is to remove the excess CO_2 from the air. Curiously, this is seldom mentioned today because "realists" have already scaled back their expectations—to merely slowing down the damage, rather than fixing the problem.

The climate scientists now say we need to stop the growth in worldwide carbon emissions before 2020, even for a compromise goal that will melt much of Greenland, flood major coastal cities, and make a third of all species extinct. (Some compromise.) Delay will take us into the territory of half of all species, failing crops, famines, mass migrations, and genocidal wars.

And the proposed treatments ought to sound familiar: we are told to walk, diet, change what we consume—that is to say, conserve energy, emphasize renewable energy, fill the car's tank much less often, and substitute nuclear-solar-wind-geothermal-hydro energy for coal. Like the diabetic who wants to avoid dying young, our civilization

1

The Big Picture

It was 1938 when the Earth's fever was first noticed. A finger was pointed at the carbon dioxide accumulating in the atmosphere from burning fossil fuels (coal, oil, and natural gas). It acts as insulation. The Earth was dressed too warmly, even then.

Now we have entered a period of consequences. Major symptoms have appeared. The climate doctors have been consulted. The lab reports have come back. Now it's time to pull together the Big Picture and discuss the treatment options.

The diagnosis, now certain, is CO_2 poisoning. We cause our planet to run a fever as we keep piling on those invisible blankets generated by cutting down forests, making cement, constantly tilling the soil, spreading fertilizer, and burning fossil fuels.

Warmer Climate on the Earth May Be Due To More Carbon Dioxide in the Air

By WALDEMAR KAEMPFFERT

According to a theory which was held half a century ago, variation in the atmosphere's carbon dioxide can account for climatic change. The theory was generally dismissed as inadequate. Dr. Gilbert Plass re-examines it in a paper which he publishes in the American Scientist and in which he summarizes conclusions that he reached after a study made with the support of the Office of Naval Research. To him the carbon dioxide theory stands up, though it may take another century of observation and measurement of temperature to confirm it [....]

The New York Times
October 28, 1956

Despite nature's way of maintaining the balance of gases the amount of carbon dioxide in the atmosphere is being artificially increased as we burn coal, oil and wood for industrial purposes. This was first pointed out by Dr. G. S. Callendar about seven years ago. Dr. Plass develops the implications.

Generated by Man

Today more carbon dioxide is being generated by man's technological processes than by volcanoes, geysers and hot springs. Every century man is increasing the carbon dioxide content of the atmosphere by 30 per cent—that is, at the rate of $1.1°$ C. in a century. [....]

Whatever the cause of the warming of the earth may be there is no doubt in Dr. Plass' mind that we must reckon with more and more industrially generated carbon dioxide. "In a few centuries," he warns, "the amount of carbon dioxide released into the atmosphere will be so large that it will have a profound effect on our climate."

Even if our coal and oil reserves will be used up in 1,000 years, seventeen times the present amount of carbon dioxide in the atmosphere must be reckoned with. The introduction of nuclear energy will not make much difference. Coal and oil are still plentiful and cheap in many parts of the world, and there is every reason to believe that both will be consumed by industry so long as it pays to do so.

Note that last sentence about coal and oil. We're now discovering the high cost of low price.

Half a century ago, scientists saw what was coming, though they underestimated how fast it would arrive because they did not sufficiently appreciate the world's appetite for fossil fuels. It's now five times greater than back then.

For the people of 2050

Geschrei

Ich fühlte das grosse Geschrei
durch die Natur

Edvard Munch, *The Scream of Nature.*
Lithograph 1895.

The volcanic eruption of Krakatoa in 1883 caused unusually intense sunsets throughout Europe in the winter of 1883–1884, which Munch captured in his painting and subsequent lithograph. The setting is a viewpoint in Oslo. The artist shows himself reacting with horror to the scream of Nature. The position in which he portrays himself is a reflex reaction typical of anyone struggling to keep out distressing noise.

Munch wrote, "I was walking along a path with two friends— the sun was setting—suddenly the sky turned blood red—I paused, feeling exhausted, and leaned on the fence—there was blood and tongues of fire above the blue-black fjord and the city— my friends walked on, and I stood there trembling with anxiety— and I sensed an infinite scream passing through nature."

Contents

needs to take all of these measures to avoid collapsing later this century.

By taxing the carbon pollution and reducing taxes elsewhere, we can make alternative energy sources the good deals—and create some real incentives to remodel buildings and buy plug-in hybrid cars.

But few ask if such measures are quick enough. Or reliable enough. Or if they can head off the developing world from repeating our mistakes.

Why should conserving energy work out any differently than the advice to eat less? Dieting really ought to work—and it does in the short run. But most dieters weigh *more* several years later. It's the same thing with stopping smoking (four out of five resume).

Do we really want to bet our only habitable planet on the success of a low-carbon diet? People may stop dieting because something stressful comes up. In human-induced climate disease, reactions to stress are also making things worse. Every summer, energy conservation backslides into burning more coal because of what happens when the air conditioning fails: people die. In buildings where the windows don't open, businesses close.

Getting another coal fix means that we spiral up, as hot begets hotter. To break this vicious cycle and restore CO_2 to normal levels, we need a treatment plan that's big enough to cover the contingencies—and fast enough to turn this situation around within several decades.

Coal is the worst of the fossil fuels, creating twice as much CO_2 as natural gas. But instead of decreasing, coal use in the U.S. is now projected to double by 2030. We're planning to build another new coal plant every month. In China, the current rate is a two new coal plants every *week*.

Because CO_2 mixes worldwide within several years and hangs around for many centuries, their CO_2 is ours and ours is theirs. The U.S. has been the world's largest contributor over the years, what with our dirty coal, long commutes, and big, boxy gas guzzlers.

An important reason to institute vigorous treatment now is that even if we stopped adding CO_2 today, delayed effects of past emissions would double our present fever by 2050.

Our window of opportunity appears to be rapidly closing. If we don't turn around emissions growth by 2020, we'll never hold the fever down enough to avoid the worst consequences. It's a catastrophe in slow motion but nonetheless a tragedy awaiting today's students.

Since we only get one shot at this time bomb, we must allow for contingencies—also rarely discussed. For example, it's quite likely that another supersized El Niño will occur in coming decades, again with major drought and fires. *But suppose it lasts twice as long as usual?* We did have a long one from 1986–87 but it wasn't also a big one.

A big, long El Niño would likely dry out two of the three major rain forests of the world. The resulting fires in Southeast Asia could inject five times the usual yearly increment of anthropogenic CO_2 into the atmosphere. If the Amazon burns off, that's an additional fifteen-year hit in only a few years.

It would cause a mass extinction of both animal and plant species, about half being lost in the aftermath.

Lacking those tropical trees to extract CO_2, the earth's fever would climb half again as fast. Forced to play catch up, we might find that we lacked maneuvering room. And crash.

So for contingencies, we must quickly create a big safety margin, above and beyond implementing the gradual improvements for the long run. Though we still have some maneuvering room, seventy years of neglect have almost painted us into a corner.

So how fast must we treat climate disease? Unjustified delay in starting treatment has happened often enough in medicine that there is now a cautionary aphorism: "The doc who waits until dead certain may wind up with a dead patient." Few climate scientists or politicians, it appears, are accustomed to thinking like physicians (or, for that matter, military officers) about the tradeoffs between urgency and uncertainty.

For global warming, the usual scientific uncertainties have been dangerously oversold by the naysayers and procrastinators. The do-nothings are like the patient who puts off treatment because the doctor isn't sure which subtype of cancer it really is. And, when that is settled, puts off chemotherapy again to shop around for "natural" treatments—then denies everything. And dies.

Rather than talk about "certainty" and the most likely climate outcome, what we need from the climate scientists and economists is a risk assessment. Risk is the likelihood multiplied by the consequences—and we have already stumbled into the high-risk zone. I'd say that we are facing a medium likelihood of widespread catastrophe, rather like flying on a plane with a 30 percent chance of losing a wing before landing.

Even though the most likely scenario is that we would arrive safely, we'd strive mightily to avoid flying on that airplane in the first place. And, presumably, work even harder to keep our kids from boarding it. But we cannot afford an endless analysis or an inconclusive debate over cost-benefit.

James Lovelock, in *The Revenge of Gaia*, is the only big thinker who seems to reflect on global warming over the next few decades like a physician thinks about the patient's situation over the next year. Lovelock says that we have to start quickly expanding a proven solution and not keep waiting for something better down the road.

When the cancer patient asks about targeted genetic treatments, the physician explains, "You don't have time to wait for them. You're in a race with a destructive process. You need something that is known to work half of the time, even though it isn't perfect and has some unpleasant side effects." The physician may silently add, "If you'd stopped smoking earlier, you wouldn't have to use this medicine now." And so it is with stopping our industrial-scale smoking and starting "chemo."

What is Lovelock's reluctant solution? In France, this carbon-free power source supplies 78 percent of their electricity. New Jersey gets 52 percent. Worldwide it has a far better safety record than any other major power source. Still, there are as many objections to nuclear power plants as there are to chemotherapy. As with chemo, there are promising improvements down the line. They appear to overcome the reasonable worries about reactor accidents, fuel diversion into nuclear weapons, and the long-term management of waste (they can even use existing waste as fuel).

But we're going to have to quickly stop smoking coal and, until something better comes along, Lovelock says that means going with what we've got, the current approved reactor designs. I'd prefer deep geothermal heat if they can ramp it up fast enough. But those are the only two routes, so far as I can see, likely to create our safety margin during the next decade.

We are remaking the earth in dangerous ways. But this should not make anyone jump to the conclusion that we are in a hopeless bind. People have proved wonderfully inventive when confronted with big challenges. For tackling ozone smog and then acid rain, there was much moaning about the price tag (industries find that useful for improving profits via tax breaks). But the estimates proved wide of the mark. Improvements in technology such as the catalytic converter greatly reduced the costs. Inventions to bring down global fever will stimulate the economy far better than the second home, the third car, or the fourth computer. Solar energy in particular will create many jobs.

Because we have to stop the growth in fossil-fuel emissions before 2020 to avoid the catastrophic consequences of a high fever, action must be swift. It is no longer possible to merely plan for the long run. But if we generate much of our electricity without fossil fuels and start driving plug-in hybrids, we can indeed make a big dent by 2020.

For people who seek meaningful work in life, the efficiency agenda starts at home and the public policy agenda won't happen without the grassroots becoming angry. The high stakes will draw even more good people into political life. Potentially it's a renaissance—though not for any country that buries its head in the sand.

Time's up. Do we, knowing full well the consequences of our inactions, really want to destroy our civilization and kill off half of all species? I trust we have more brains than that.

[N]umerous long-term changes in climate have been observed. These include changes in Arctic temperatures and ice, widespread changes in precipitation amounts, ocean salinity, wind patterns and aspects of extreme weather including droughts, heavy precipitation, heat waves and the intensity of tropical cyclones.
—Intergovernmental Panel on Climate Change (IPCC), 2007

[F]or the past twenty years, the period during which greenhouse science emerged, most of the effects of heating on the physical world have in fact been more dire than originally predicted.

The regular reader of *Science* and *Nature* is treated to an almost weekly load of apocalyptic data, virtually all of it showing results at the very upper end of the ranges predicted by climate models, or beyond them altogether.

Compared with the original models of a few years ago, ice is melting faster; forest soils are giving up more carbon as they warm; storms are increasing much more quickly in number and size.
—author Bill McKibben, 2006

The future holds more frequent episodes of violent weather.

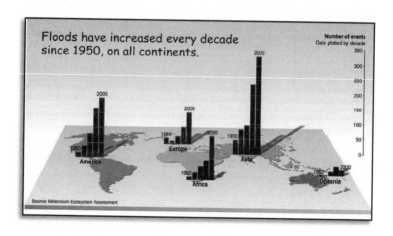

Indeed, it started happening a half century ago.

2

We're Not in Kansas Anymore

Does the climate news seem like a jumble of issues with few obvious connections to tie them together? People are broiling (35,000 Europeans died in the 2003 heat wave). Greenland is acting like a runaway ice-cube machine. Ski season is shrinking. Coral reefs are bleaching, sea level is rising. We experience stronger hurricanes, more wildfires, and new insects arriving in the neighborhood.

At least those things hang together under the rubric of "It's getting warmer." But our present climate change isn't a simple one-dimensional problem that can be framed as mere warming, what with floods and droughts becoming much more frequent, deserts expanding, and the like.

Then there are varied causes such as the CO_2 accumulating, the sun brightening, the ozone thinning, the methane soaring, the Gulf Stream slowing, the lower atmosphere thickening, and El Niño possibly settling in for good. All of that tipping, slipping, and flipping. Where do they fit into the big picture?

South of Japan in 2006, there were three tropical cyclones
(also known as hurricanes and typhoons) at once.

My position in your chain of information is analogous
to that of your primary-care doctor, reporting on the
results of the tests and the analysis of the specialists, trying
to put together the big picture, and helping you to
understand the treatment options.

Climate involves a Rube Goldberg chain of knock-on
effects, not unlike the everyday chain of events in our
bodies that biologists study. Medicines typically intervene

at one link or another of a long chain or web. And so we may find a number of places to intervene in climate disease. I'll use my perspective on a century of medical progress to show why I'm optimistic that climate science will lead to effective climate medicine.

It is becoming urgent that you understand much of this because, just since the twenty-first century began, climate change seems to have jumped to the fast track. There are now swarms of summertime-only earthquakes coming from the exact places in Greenland where outlet glaciers come down to the sea and produce icebergs. When the number of quakes doubles, and then redoubles—all in only ten years—it's like feeling the ground shift under your feet.

It seems that we're not on the slow track anymore. For many of us, it was a moment like that in the *Wizard of Oz* when Dorothy says to her dog, "Toto, I've got a feeling we're not in Kansas anymore."

Anyone who reads some history, classics, or anthropology knows that past civilizations have proven fragile. But then past societies did not have very many scientists or historians, who sometimes develop a perspective on a problem that has major implications for what society must do next.

In his excellent book *Collapse: How Societies Choose to Fail or Succeed*, Jared Diamond takes us through some societies that crashed (and a few that pulled back from the brink), just as his earlier *Guns, Germs, and Steel* took us through the biogeography that helps to make a society a winner.

Many of the dozen factors that Diamond analyzed in *Collapse* are things that gradually creep onto the scene, such as overpopulation, deforestation, contamination, and the soil washing out to sea in muddy rivers. However, past societies "declined rapidly after reaching peak numbers and power, and those rapid declines must have come as a surprise and shock to their citizens." Gradual creep makes you think in terms of a gradual decline, not the rapid collapse that Diamond reports. Creeps can suddenly turn ugly.

We are living in a house of cards and we'd better understand its vulnerabilities in order to strengthen its foundations. But a disaster of biblical proportions isn't inevitable. There are quite a number of ways to clean up our act and then clean up the excess carbon in the air. I discuss them starting at Chapter Sixteen. If in despair, skip ahead.

Climate change is a moral issue as well as a political one. Though we have turned a blind eye on our invisible pollution of others, the United States has risen to ethical challenges many times in the past, from abolishing slavery, to women voting, to civil rights. It would now be appropriate for the U.S. to take the lead in replacing coal-fired plants, doubling fuel efficiency for cars, and providing clean power and vehicles for the developing countries.

We must get our political leadership to pay attention before our society becomes too weak to move effectively.

August 1941

August 2004

Muir Glacier, Alaska

Suppose that we're already seeing the consequences of global fever? And that it's been going on for at least a half century?

As temperature rises in the western USA, there are more fires.

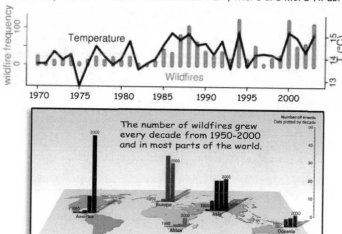

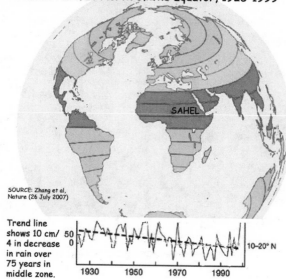

Reduced Rainfall North of the Equator, 1925-1999

SOURCE: Zhang et al,
Nature (26 July 2007)

Trend line
shows 10 cm/
4 in decrease
in rain over
75 years in
middle zone.

Overall, the risk of sea-level rise from global warming is less at almost any given location than that from other causes, such as tectonic motions of the earth's surface.
>—climate scientist (and contrarian) Richard Lindzen, 2007

If 98 doctors say my son is ill and needs medication and two say 'No, he doesn't, he is fine,' I will go with the 98. It's common sense—the same with global warming. We go with the majority, the large majority... The key thing now is that since we know this industrial age has created it, let's get our act together and do everything we can to roll it back.
>—Governor Arnold Schwarzenegger of California, 2007

Essentially all of the observed climate-change phenomena are consistent with the predictions of climate science for greenhouse-gas-induced warming. No alternative "culprit" identified so far—no potential cause of climate change other than greenhouse gases—yields this "fingerprint" match.

A credible skeptic would need to explain both what the alternative cause of the observed changes is—*and* how it could be that greenhouse gases are not having the effects that all current scientific understanding says they should have. No skeptic has done either thing.
>—climate scientist John Holdren, 2006

There is a small chance that the skeptics are right, or we might be saved by an unexpected event such as a series of volcanic eruptions severe enough to block out sunlight and so cool the Earth.

But only losers would bet their lives on such poor odds. Whatever doubts there are about future climates, there are no doubts that both greenhouse gases and temperatures are rising.
>—physiologist James Lovelock, 2006

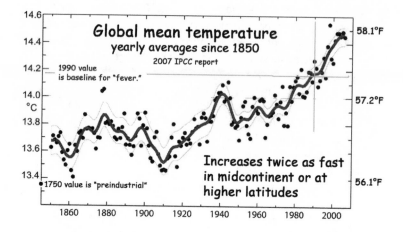

Global mean temperature
yearly averages since 1850
2007 IPCC report

1990 value is baseline for "fever."

1750 value is "preindustrial"

Increases twice as fast in midcontinent or at higher latitudes

Global mean temperature is calculated from averaging the air temperatures at 2 m above the land or sea surface. Since the oceans have evaporative cooling, the warming has been greater over land. Oceans are also 70 percent of the total surface area.

Most temperatures in this book are temperature differences, not the thermometer reading. Just as a physician might talk of two degrees above normal body temperature as "2° of fever," so climate scientists speak of "temperature anomalies."

What, however, is the equivalent of normal temperature for Mother Earth, the baseline from which anomalies depart? This is arbitrary, but usually it is the global mean temperature about 1990 (actually the average of 1980–1999 readings).

The other baseline temperature used is the preindustrial temperature, usualy that estimated for 1750. It is 0.8°C below the other. So when we talk of the prospects for a 3° fever this century, it's 3.8°C above that at the time when our fossil-fuel fiasco began.

The slow temperature rise between 1950 and 1975 confused climate scientists at the time. It now seems likely that vehicular smog played a major global dimming role (Chapter Nine) until the catalytic converter made a difference after 1975. U.S. oil tripled in those years while coal was unchanged (see page 26).

3

Will This Overheated Frog Move?

Particularly in a country of immigrants like the United States, there is a pattern that you see over and over. Parents work very hard at creating a better world for their children, settling for less for themselves if it provides better prospects for their children.

Writ large, it's the story of constructing civilization. Each generation tries to leave things better for the next.

Sometimes things go downhill instead. Indeed, most societies have failed eventually. Often it has been a matter of fouling the nest—as when irrigation eventually adds so much salt to the soil that crops fail. Likely they had no idea what was wrong and thus no idea how to fix things.

One of the sad things about our present situation for my generation is that we seem likely to leave the earth in worse shape than when we took over running things. And this time, it's not because of ignorance. Thanks to science and history, we've known for a long time the conseq-uences of our actions—and kept doing them. The people building more coal-fired plants have long known what they are doing to the health of the people downwind, even if they didn't understand the greenhouse implications.

The nineteenth-century physicists Joseph Fourier and John Tyndall realized that gases in the earth's atmosphere trapped heat. Then in 1898, Svante Arrhenius in Stockholm calculated that a doubling of the pre-industrial amount of carbon dioxide in the air would cause the Earth to run a fever of $5^{\circ}C/9^{\circ}F$.

Forty years after Arrhenius, Guy Callendar in the UK noted that the temperature had risen and that the CO_2 was up by at least 10 percent.

In 1956, the American physicist Gilbert Plass got good data on how infrared radiation gets trapped by the atmosphere and began cautioning that rapid industrialization could produce serious climate change.

The oceanographers led by Roger Reville reported that the oceans just were not up to the task of taking the fossil-fuel CO_2 out of the air in time. In 1958, Dave Keeling started making monthly measurements of CO_2 atop Mauna Loa and confirmed the upward trend with very precise data.

So it's been over half a century since the scientific community realized that greenhouse warming was actually underway and started warning the politicians and the public. Soon thereafter, we began to see some concerted, well-funded efforts to confuse you about the subject. I'm not talking here of disagreements among scientists but of disinformation campaigns to delay action and profit by it. This manufacturing of controversy included such seductive mantras as "scientists are still uncertain"—as if that excused everyone from doing anything inconvenient this year.

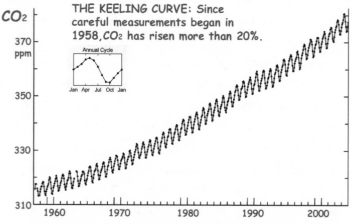

You can actually see the earth breathing in the seasonal variations of the Keeling Curve. The northern forests take some CO_2 out of the air each summer via photosynthesis, cutting into the upward trend. The CO_2 is measured every month in the trade winds flowing over the top of Mauna Loa, Hawaii, well away from any local sources that might complicate the interpretation. By 1961, it was clear there was an upward trend. That gap in 1964 is because the project ran out of money. The Keeling curve has now become twice as steep. With business-as-usual use of fossil fuels, it should climb even faster.

People accustomed to managing high-risk situations—say, my physician colleagues in the medical school—were appalled by the emphasis on "certainty" invented by the disinformation industry. You're not going to get certainty about climate illness any more than you do with human disease (or career choices, for that matter). And you can't delay acting just because of remaining uncertainty.

There's a clock running but you don't know when the tipping point will arrive and the patient will slip into irreversible damage. You often have to start treatment early for the most likely disease while continuing diagnostic tests.

No democracy is going to tax itself for a remake without a wide acceptance of what is at stake. Most of us are not climate scientists but many of us will need to understand what's up if we are to collectively take effective action—and we will need to do this on the fly, without going back to school or leaving the problems for the next generation to handle. Our civilization is going to have to make some serious decisions quite soon or risk participating in the familiar downhill spiral that leads to the collapse of a civilization, with ours joining dozens of prior societies in the dust bins of history. Flushed.

That could happen within the lives of people you know, such as you and your children. (I used to say "your children and grandchildren," but things have been speeding up.)

Let me recast that apocryphal story about cooking a frog. If a frog jumped into a steaming hot tub, he'd leap back out. But if he was already sitting in the tub's rim at the time that the heater was turned on, he might not notice the rising temperature until he had become too weak to respond.

We're not that lethargic frog, however. Frogs don't have foresight and we do. Furthermore, thanks to science, our civilization has developed a warning system. And thanks to our technology, we may be able to rescue ourselves.

Unfortunately, fifty years of warnings have been ineffective at moving most politicians. We'll cook with them if we don't convert them or replace them.

The debate about global fever per se is over. Yes, most of the Earth is overheating. That part is getting a little difficult to deny.

And, yes, it's mostly caused by human actions and inactions. A half century ago, that was still an open question. Now it's clear that the addition of CO_2 to the atmosphere is responsible for much of the overheating after 1950. Methane contributes another big chunk.

Past greenhouse episodes have been due to methane burps, the prolonged belching of volcanoes, or, most recently, the earth getting closest to the sun in July when the tilt of the earth's axis was greater than it is today. This time we did it all by ourselves, via low tech means of burning fossil fuels, making cement, and clearing land.

Europe tackled the problem directly and has markedly reduced its use of fossil fuels since 1980. But an American now uses twice as much energy as a European (and throws out twice as much garbage, and uses twice as much toilet paper).

What's debated now is what happens next. We have good models of future climate. And we have good political models, such as the 1987 Montreal treaty that reversed the trend in the ozone hole by changing industrial chemicals; the hole itself may be healed by midcentury. It's a great success story of how science and industry, politicians and diplomats can cooperate to reverse a big problem in the making.

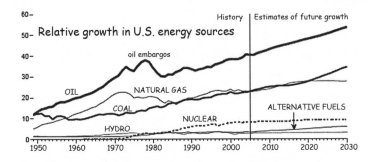

It's worth noting the differences between ozone then and fossil fuels now. The most obvious is that the chlorofluorocarbons (CFCs) were produced by relatively few manufacturers, who could easily find substitutes. The public could hurry them along, since we could readily avoid many products containing CFCs. Fossil fuels are big business worldwide, however, with lots invested up front for profits decades later.

So let us look at the role played by the scientists, the media, special interests, the public, and governments. I'll elaborate on the perceptive comments of climate scientist Jim Hansen to the U.S. National Academy of Sciences.

For ozone and CFCs, scientists transmitted a clear warning—while for global fever, scientists didn't make a clear distinction between generic climate change and a major makeover, a wholly different planet to live on.

A climate success story

1987 Montreal Protocol signed

million tons of CFC

1974 discovery of what CFCs were doing to the ozone layer

industry took note, stopped expanding uses

100,000

10,000

CFCs were widely used in spray cans, air conditioners, refrigerators, and for making styrofoam. Production increased 10% each year.

1,000
1940　1950　1960　1970　1980　1990　2000

The media transmitted the ozone story well, but for global fever, they got suckered into always "balancing" the science, often with disinformation. Then they leaped to hopelessness, complete with sports metaphors.

The special interests for CFCs denied the science for years but stopped building new capacity and pursued technological innovation. But for global fever, some of the biggest of the oil and coal companies live in their own little bubble of unreality and pursue dangerously irresponsible disinformation campaigns analogous to the misleading representations that constitute fraud in other arenas.

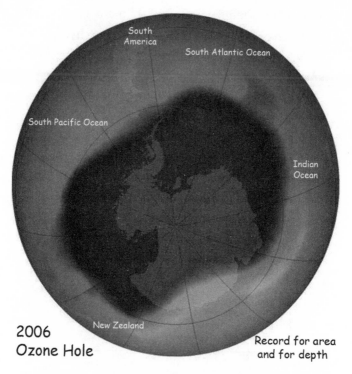

2006
Ozone Hole

Record for area
and for depth

Looking down on the South Pole at the record-sized ozone hole of 2006. It should gradually become smaller as the CFCs decay.

The public had a quick response to the ozone warning. The marketplace worked: the popular spray cans became unpopular and no more CFC infrastructure was built. With global fever, the U.S. public is understandably confused and in 2005 ranked environmental problems low on its list of concerns (and, within environmental problems, ranked global warming near the bottom). Things have since changed.

Governments, with leadership from the U.S. and Europe, did well on ozone and, though it took thirteen years, the Montreal Protocol was finally signed and, two years later under the leadership of Margaret Thatcher, it was considerably strengthened. Europe has taken the lead for the last quarter century on global fever issues, with France getting 78 percent of its electrical power from nuclear reactors with zero carbon emissions.

The first rule of kindergarten remains "Clean up your own mess." Yet the U.S. government fails to lead, and seems overwhelmed by the special interests. It has taken the lead in delaying effective action with its quibbles about the Kyoto Protocol, with obstruction and foot-dragging tactics at every climate conference since 1997.

The U.S. ought to be using its technological proficiency to solve the CO_2 problems, creating a good example for developing countries to follow, rather than setting them up to feel that their accomplishments will be dwarfed by the profligate waste of the world's leading polluter.

We scientists have also been pulling our punches, giving out "climate lite" when we should have been stating the prospects as we saw them, "proven" or not. George Monbiot gives a good example:

> At a meeting of climate change specialists, Sir David King [chief science advisor to the British government] announced that a "reasonable" target for stabilising carbon dioxide in the atmosphere was 550 parts of the gas per million parts of air. It would be "politically unrealistic," he said, to demand anything lower.

Simon Retallack, from the Institute for Public Policy Research, stood up and reminded Sir David what his job was. As chief scientist, his duty is not to represent political reality—there are plenty of advisers schooled in that art—but to represent scientific reality . . .

Sir David replied that if he recommended a lower limit, he would lose credibility with the government. As far as I was concerned, his credibility had just disappeared without trace. By shielding his masters from uncomfortable realities, he is failing in his duties as both scientist and adviser.

It's one thing for politicians to engage in understatement when trying to build a consensus or to avoid a second controversial subject when addressing the first—I suspect Al Gore of doing this in *An Inconvenient Truth* when he leaves nuclear power off his list of wedges—but we scientists need to be more straightforward in speaking to the people who need to get their information unfiltered, with the sugar coatings left off.

Climate change, and how we address this issue, is a defining issue of our era.

—UN Secretary-General Ban Ki-moon, 2007

What we need now is good information and careful thinking, because in the years to come this issue will dwarf all the others combined. It will become the *only* issue.

—biologist Tim Flannery, 2005

When I give talks on global warming, quite a few of my over-50 peers in the audience remark that this is, at its heart, an issue of legacy. It is our children's climate, and our grandchildren's, that is being shaped by the building greenhouse effect. One disturbing part of that legacy is this: while half the gas billowing from smokestacks and tailpipes is typically absorbed by the oceans or plants each year, the rest remains stashed in the air for a century or longer, building like unpaid credit card debt.

—science writer Andrew Revkin, 2007

Governments... spend a small slice of tax revenue on keeping standing armies, not because they think their countries are in imminent danger of invasion but because, if it happened, the consequences would be catastrophic. Individuals do so too. They spend a little of their incomes on household insurance not because they think their homes are likely to be torched next week but because, if it happened, the results would be disastrous. Similarly, a growing body of scientific evidence suggests that the risk of a climatic catastrophe is high enough for the world to spend a small proportion of its income trying to prevent one from happening...

The real difficulty is political. Climate change is one of the hardest policy problems the world has ever faced. Because it is global, it is in every country's interests to get every other country to bear the burden of tackling it. Because it is long term, it is in every generation's interests to shirk the responsibility and shift it onto the next one. And that way, nothing will be done...

Developing countries argue, quite reasonably, that, since the rich world created the problem, it must take the lead in solving it. So, if America continues to refuse to do anything to control its emissions, developing countries won't do anything about theirs. If America takes action, they just might.

— from the *Economist*, 9 September 2006

Our understanding of the Earth system is not much better than a nineteenth-century physician's understanding of a patient. But we are sufficiently aware of the physiology of the Earth to realize the severity of its illness.

We suspect the existence of a threshold, set by the temperature or the level of carbon dioxide in the air; once this is passed nothing the nations of the world do will alter the outcome and the Earth will move irreversibly to a new hot state.

We are now approaching one of these tipping points, and our future is like that of the passengers on a small pleasure boat sailing quietly above the Niagara Falls, not knowing that the engines are about to fail.

—physiologist James Lovelock, 2006

The widespread confusion about our climate crisis is no accident. For more than a decade, those who deny that climate change is an urgent problem have sought to delay action on global warming by running a brilliant rhetorical campaign and spreading multiple myths that misinform debate. As a result, many people still believe global warming is nothing more than a natural climate cycle that humans cannot influence, or that it might even have positive benefits for this nation. Neither is true. The science is crystal clear: We humans are the primary cause of global warming, and we face a bleak future if we fail to act quickly.

—oceanographer Joseph Romm, 2007

We never have 100 percent certainty. We never have it. If you wait until you have 100 percent certainty, something bad is going to happen on the battlefield.

—General Gordon R. Sullivan, 2007
(former Chief of Staff, U.S. Army)

4

"Pop!" Goes the Climate

I remained comfortable with the slow-but-sure metaphors for climate change until one day in 1984 when I learned that things could go "Pop!" That's when, quite by accident, I heard about the surprises from the past seen in the ice cores pulled up from the depths of Greenland's ice sheet.

The ancient climate records were full of sudden jumps in temperature and rainfall, lasting many centuries before flipping back. In the course of five minutes into a lecture by a visiting Swiss geochemist, Hans Oeschger, I had to abandon my limited notions of gradual change. Flips occur. They are commonplace events.

It was certainly unsettling—and even more so as I began to think about it. Anything that abrupt is probably not a simple analog tweaking at work—say, the sun getting hotter or cooler—but suggests there is a big positive feedback loop hidden somewhere.

Cause and effect get all mixed up when there's a feedback loop involved. You really have to think about it differently, as reactions can be all out of proportion to the stimulus. We call a person "excitable" if a creeping pint-size stimulus suddenly elicits a gallon-sized response. We neurophysiologists study nerve and muscle cells with positive feedback mechanisms that help things to happen very quickly. With them, things often go "Pop!"

Often climate change is not, in the manner of a dimmer switch, proportional to the provocation. It is more like the traditional light switch which, with slightly more pressure, suddenly blinds you in the middle of the night.

Remember those cameras with an automatic pop-up flash, that flipped up right in front of your other eye when you were about to take a picture in insufficient light? Some people, startled, dropped their new camera. We have, it now appears, a pop-up climate to prepare for.

Climate abounds with situations where, even without a constant push from the outside, the internal dynamics propel change. Collapse. Runaways. Take offs. Some tipping points are especially serious because of the demolition derby that follows: Say, drought causing the Amazon rain forest to burn off (discussed in chapter 13).

To place all of this discontinuity in perspective, consider some less dramatic everyday examples. Processes often have "modes" of operation, such as gallop and walk. You pop from one to another without an intermediary mix.

Matter often comes in "states." Good old H_2O can be found in three states: as a solid called ice, as a liquid called water, or as a gas called water vapor (which we feel as humidity). We can go back and forth between states, even jump over the liquid state via sublimation of the ice directly into water vapor. Thawing and evaporation are reversible by freezing and condensation.

Many climate processes are somewhat reversible. But it takes forever. Though we are still trying to digest the news about the slow track, some aspects of climate change may have already jumped to the fast track. Major glaciers in both Greenland and Antarctica are now dumping far more ice into the oceans than they were just five or ten years ago.

The biggest source of our climate problem is the soot and carbon dioxide from burning oil and coal. This has created an unwelcome blanket of pollution around the Earth. The global fever that results, especially in the high latitudes, is already causing sea level to rise and the flushing of the North Atlantic to diminish. Climate scientists are now predicting permanent flooding of countries such as the Netherlands, Bangladesh, Tuvalu, and the Maldives, not to mention such low-lying cities as London, Boston, New York, Miami and New Orleans.

In midcentury, about 70 percent of us will still be around, plus the kids and grandkids born between now and then. Historians will be busy writing books about the events in the hundred years since the first deadly serious greenhouse-is-happening warnings. Readers will be all too familiar with what did happen, but curious about why and how it all happened.

"Knowing what they did, how could they have done this to us" may be a familiar refrain. "They kept saying that the chances of climate change were low—but even if they had been right about that, to ignore the huge potential consequences was really stupid." Risk is the chance an event will happen, multiplied by the consequences if it does happen. But this doesn't register with many people. It took fifty years to largely overcome the reluctance to wear one's seat belt.

Historians will have a more nuanced view and one of their themes will be how fear of nuclear power led to the vast expansion of the dirtiest power source of all, coal. "The Green emphasis on small individual efficiencies worked about as well as dieting worked to keep weight off," some historian might say.

Others will focus on the disinformation campaigns created by Big Coal and Big Oil to maintain business-as-usual profits. And why they succeeded better in some countries than in others.

They will all analyze how big decisions were made (or avoided) by politicians and regulatory agencies. And how normal scientific uncertainty (the stance we routinely adopt regarding a problem that is incompletely solved) was exploited by them all in order to postpone inconvenient decisions.

Still, anyone who spots a burning house has a civic and moral duty to spread the alarm and awaken the occupants. But before acting, how certain must you be? Can you tell the difference between smoke and the exhalations of the clothes dryer? Better safe than sorry. Many climate scientists, historians will note, faced this dilemma squarely and issued warnings.

The universal lament in 2050, however, may be that the problem proved too big and complicated for nonscientists to comprehend.

That's not my experience so far. Even lawyers and policy types—and others who might have earlier skimped on science courses—are perfectly capable of appreciating how the climate story hangs together.

While I'm a medical school professor and not a climate scientist, I've been following the climate story for a quarter century by now, and trying to explain it to general readers for a decade. (I wrote the first major magazine article on the climate flips back in 1998, a cover story for the *Atlantic Monthly* called "The great climate flip-flop.")

My experience is that good analogies from everyday experience suffice to get across the main ideas of how one thing leads to another—and where we might intervene to break up feedback loops.

This book is not the place to get the latest survey of the latest climate data. Much of who, what, when, and where is always a little out of date. My book is more about the principles of acceleration—which probably won't change—that underlie tip, slip, and flip. But why and how are what you need for a deeper understanding of why we're in this mess and how we might intervene.

Larsen B ice shelf collapse 2002

Clouds
distort

Opposite: The Antarctic Peninsula is the most rapidly warming place on earth. Over several weeks of 2002, an ice shelf there, bigger than Rhode Island, was observed to shatter. This made room for more ice to flow downhill and raise sea level.

An ice stream well uphill of Larsen B has been observed to stop advancing during low tide but, as the tide rises and lifts the terminus, it accelerates to 1 m per hour.

The worry expressed then was that the large West Antarctic Ice Sheet could do the same thing if summer temperatures there were to rise above freezing. That occurred in the summer of 2005 when a surface area as large as California melted and then refroze.

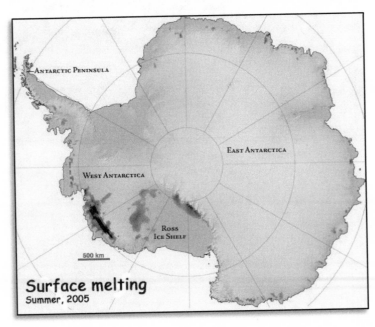

Surface melting
Summer, 2005

Oklahoma farm yard in 1936, deep dust covering fences.

5

Drought's Slippery Slope

Suppose that I told you that, even without global fever, the chance of twenty-first-century America suffering a century-long, widespread drought is one in four. Speculation? No, such chances are what history tells us—the type of history written in thick or thin tree rings.

Even without global fever's rearrangements, the Great Plains and the West have often suffered century-long droughts, far more widespread and long-lasting than the more familiar Dust Bowl of 1932–38. Droughts provide a common setup for famine, so let me show how the slippery slope works in a regional drought. Drought also illustrates the role of random events in what is otherwise deterministic.

Drought watchers define drought as a moisture deficit severe enough to have social, environmental, or economic effects. That's a *what*-based definition, and it often suffices for lawyers and policy types.

I prefer to focus on *how* a drought gets going because it shows how slippery slopes are created. And how you can "get stuck" in a drought, seemingly unable to pop out.

Slipping on the proverbial banana peel is bad enough. Slippery slopes, however, are where things get bad and, because of that, automatically get worse. This repeats and repeats, creating an increasingly rapid descent into hell. Or thereabouts. Take, for example, the time that I slipped on an unseen layer of thin ice outside the back door. Quick reflexes didn't save me, and so I sat down hard. Then, astonished, I bounced down two flights of slippery steps, gaining speed with each additional bump.

That's similar to what goes on in a drought: a moisture omission that achieves a slippery slope, a chain of events that kills off plant life, roots and all, forcing it to start anew. It's like a computer crash and reboot. The crash is fast, the reboot is slow. So here's my slippery slope view of drought. It's more about the how—the underlying mechanisms—than it is about the what.

Usually by chance, the storm tracks skip over a region for several years running. So the topsoil gets a little dry and plants grow slowly. And then a few weeks of really hot weather bakes all of the ground that isn't shaded.

The result is that leaves droop. Evaporation from the leaves is what pulls more water up from the roots, what makes a leaf fill out and the plant stand upright. No more water around the roots, and the leaves wilt. Result: no more shade.

And now the feedback: The ground which was still shaded now bakes as well. The plant may no longer stand at attention (indeed, fire may prune it). The topsoil gets much drier. This, by itself, does not seem to have a slippery slope: once the shade is reduced to nothing, it cannot get less shady.

But now another mechanism kicks in. Over a tropical forest, about half of the rainfall comes from what recently evaporated from the leaves upwind.

Less evaporation diminishes future rainfall by reducing the humidity. Think of it as recycling or, better yet, as priming the pump. To get mere humidity (water vapor) to coalesce into the little water droplets that we see suspended in steam and clouds, it takes a certain minimum amount of water vapor. This varies with temperature, measured as the dew point. Such droplets may condense to form the larger drops, finally becoming too heavy for buoyancy to keep them aloft. In short, it rains.

So the plant-generated humidity helps to harvest any passing water vapor by boosting it up into raindrop mode. It builds up the clouds. No evaporation means less rain. Once again, things get worse in a drought, almost automatically.

Now the water table drops significantly, and so the roots can't reach it. No plants grow the next spring. Surviving plants may now die, roots and all. Then, what rain that does fall is likely to run off quickly, since dead plants no longer extract water from the topsoil.

There's even more to drought's slippery slope. Dry out the ground and it gets a hard crust, thanks to the wind removing the softer bits. Thus when rain does fall, it splashes. Most runs off sideways on the hard surface rather than sinking into the soil. And so the water table drops even further.

This can be reversed by enough rain, but it takes more than one good year for storm tracks.

This chain of events can be created, on a much faster timescale, in your own back yard. You may not be able to activate all of the feedback loops by failing to water the garden all summer, but you can see most of the chain except for evaporative rainmaking.

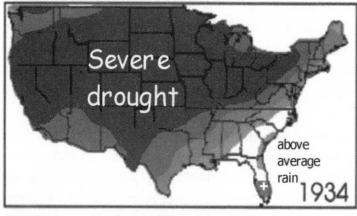

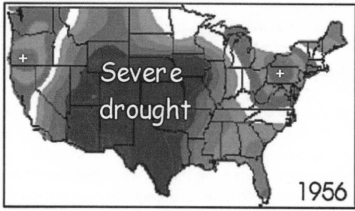

River sediments, tree-ring growth records, and the bathtub ring around lakes are the easy measures of earlier droughts, and sometimes the archaeologists can add the dates when settlements are abandoned. On top of that, you want to know the territory occupied by the drought and typical durations.

> People who remember the 1930s Dust Bowl might think they have seen the worst drought nature can offer. In the toughest Dust Bowl years, between 1934 and 1940, millions of acres of Great Plains topsoil blew away in colossal dust storms...Hundreds of thousands of people, including 85 per cent of Oklahoma's entire population, left the land and trekked west. All it took was an average 25 per cent reduction in rainfall.
> —the writer Mark Lynas, 2007

The Dust Bowl drought was before my time, but my mother said that when it did rain in Kansas City, it rained mud.

The Dust Bowl affected much of the western half of the U.S, magnifying the effects of the economic depression that began several years earlier when a speculative bubble crashed the stock market. In 1934, almost the entire United States was in severe drought.

And it wasn't as if the rain fell elsewhere. Only southern Florida got more rain than usual. In contrast, the big drought of 1956 was associated with extra rainfall in the Northwest and Northeast.

The computer models of climate (more in chapter 15) give some cause-and-effect insight into the rainfall patterns of 1934. If you force the model to follow the

observed temperatures in the Pacific Ocean—cool since a
La Niña was in progress—it produces a rainfall map that
looks very much like the actual 1934 rainfall map.

The previous decade-scale drought was back in the
1890s. There was another in the 1850s, just before the U.S.
Civil War, and another back in the 1660s, when Isaac
Newton was contemplating falling apples. They too
covered a lot of territory at once. Three of the four
happened during the warm up after the Little Ice Age
ended about 1850.

Five widespread decade-scale droughts in the last 400 years

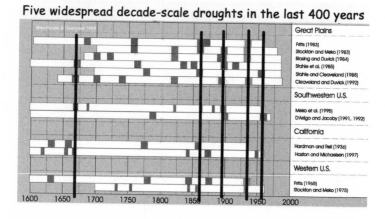

Notable droughts (dark boxes) west of the Mississippi River in the
last 400 years. Of the five widespread decade-scale droughts
(connected by vertical lines) comparable to the Dust Bowl, four
have occurred since the earth began warming up after the Little
Ice Age ended in about 1850. (Data compiled by Woodhouse and
Overpeck 1998. The light horizontal ribbon represents the time
span of the data at the location.)

Humans, of course, can make natural drought cycles worse. The longer I hang around a medical school, for instance, the more stories I hear about common patient behaviors that just exacerbate the problem.

There's a big market out there for denial. "This is not happening to me" is one stage of grief. I hope we don't see the other two: anger ("How dare God do this to me") and bargaining ("Just let me live to see my daughter get married"). Some remind me of what seems to be going on with the Earth's fever.

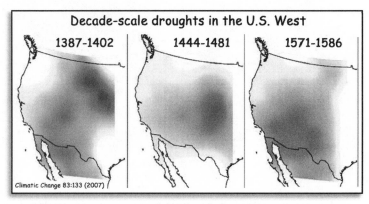

All of these Little Ice Age droughts in the American West were worse than the Dust Bowl of the 1930s, even though there were fewer century-scale droughts then.

With global warming will come additional denial, certainly by some of the naysayers and procrastinators of the past. While denial has been encouraged by disinformation campaigns, it's already common enough for climate problems. Take the 1930s Dust Bowl. Drought wasn't the only cause of the black blizzards. Not only had the farmers done away with the grasses that held the topsoil in place, but growing wheat had exhausted the topsoil. Overgrazing by cattle and sheep had stripped the soil further. So when the rains failed, the land just blew away.

Dusty dune formed after 1935 dust storm.

Starting in 1935, federal conservation programs tried to change the business-as-usual farming practices by seeding grass, rotating crops, strip and contour plowing, and planting trees as windbreaks. But the farmers were in denial about their traditional careless practices. For many of them, it took the incentive of the government paying them a dollar an acre to get them to grudgingly adopt the new methods.

The denial went deeper than that, reflecting something about national traits as well. As the historian Robert

Worster wrote, "The ultimate meaning of the dust storms of the 1930s was that America as a whole, not just the plains, was badly out of balance with its natural environment. Unbounded optimism about the future, careless disregard of nature's limits and uncertainties, uncritical faith in Providence, devotion to self-aggrandizement—all these were national as well as regional characteristics."

So anyone who would tackle our current addiction to fossil fuels is going to have to maneuver around denial. But there is another not uncommon patient behavior to bear in mind as well: aversion to experts. In medicine, "too late" is often heard. In many parts of the world, that's because of the poor availability of medical care. Where that isn't the case, you still see the patient who overgeneralizes about herbs, assuming they can fix anything. Or the patient who finally sees a nurse-practitioner but resists referral to a specialist because "I just don't feel comfortable with fancy specialists."

There are a lot of people out there like that, and I fear they will react to climate experts in a similar way. Even if education might eventually reduce such problems, climate may require much quicker action than several generations of education. The urgency about climate change really calls for politicians and cultural leaders to interpret the warnings. They will be heard in a way that the climate experts may not.

It's called leadership.

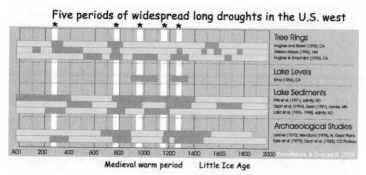

Five periods of widespread long droughts in the U.S. west

Tree Rings
Lake Levels
Lake Sediments
Archaeological Studies

Medieval warm period   Little Ice Age

Century-scale droughts (*gray boxes*; the more brief droughts are omitted here) over the last 2,000 years, west of the Mississippi River. The ones that were widespread are connected by the *white vertical bars*. There were five such widespread century-scale droughts in the last twenty centuries, but four of them occurred in the span of the Medieval Warm Period. None occurred in the Little Ice Age that followed.

We have obviously survived a number of climate changes already. Some of them even sped up the evolution of *Homo sapiens,* as I discussed in *A Brain for All Seasons.*

However, 99 percent of Americans no longer live on farms and are instead crowded into cities and their suburbs. Most are unable to feed themselves without food on the shelves at the local grocery. This inherently unstable situation makes us much more vulnerable. When I was born in 1939, 20 percent of Americans still lived on farms. Now it's 1 percent.

So what are the odds of collapse from climate problems? We naturally focus on the next one, when it comes to natural disasters, but you also have to think about how often something happens.

Besides floods, we can give the odds of drought conditions. We now have a lot of data on past droughts. Some
areas, such as present-day North Dakota, stayed in
drought for 700 years at a time. So let's focus on century-
scale droughts (say, lasting at least ten times as long as the
Dust Bowl). Furthermore, let us restrict ourselves to the
occasions when most of the western half of the U.S. was
simultaneously in drought conditions.

Such a drought (shown as a vertical white line), covering the same territory as the Dust Bowl but lasting a
century, has indeed happened. If it happened again, it
would be a disaster, and not only for the farmers. Even if
famine were not our fate, a country weakened by such a
problem may be taken over by some other country—making drought a serious national security concern. Not only
did such a widespread, century-scale drought happen
once but it has happened five times in twenty centuries.
One century in every four, on average.

That track record for the western half of the U.S., even
without global fever, suggests that it has a 25 percent
chance of getting hit by a thoroughly disastrous century-
scale drought in the twenty-first century. If you count less
widespread droughts or ones that last less than fifty years,
the chances of serious disruption are much higher.

The now-revealed unstable climate lacked our current
rate of global warming. We can get some clues about the
consequences of warming, however, from when the five
century-scale droughts occurred.

Though there were serious decade-scale droughts (page 47), no century-sized droughts occurred in the U.S. during the Little Ice Age (from roughly 1315 to 1865) when Europe and some other parts of the world were generally about 1°C cooler.

Four of the five long American droughts during the last twenty centuries happened during the so-called Medieval Warm Period of anomalous climate.

So instead of one century in four, perhaps our chances are closer to four chances in five of the twenty-first century having a major climate disturbance as some regions warm more than others. An 80 percent chance of having to live in a vastly disrupted United States is so close to a sure thing that Americans ought to insist that their government treat it that way as a precaution.

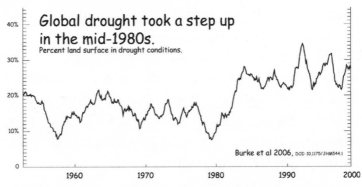

Global drought took a step up in the mid-1980s.
Percent land surface in drought conditions.

Burke et al 2006, DOI: 10.1175/JHM544.1

Fifty years ago, climate scientists didn't know most of this history and didn't know much about the feedback loops. Thanks to science, we now know the chances. Furthermore, we know the mechanisms that create the drought's slippery slope, and that may someday help us to intervene and limit the damage.

Wall of dust in April 1935 approaching Stratford, Texas.

No one seems to know this American drought history, despite the media's well-known appetite for disaster stories. And fewer people realize what is arriving on climate's fast track. Most of the models agree in predicting that the dryness of the 1930s Dust Bowl will return to the American Southwest by midcentury—and for good.

Yet it's business as usual for most of our leaders. While they should of course be better informed, it's really we citizens who need to know this history and, at the least, some metaphors for how it happens. Without our express-ed concerns, politicians will continue to "study the prob-lem further," not realizing the dangers of the slippery slope.

Is anyone concerned with public policy (say, a presidential commission) looking ahead at the major disruption problem? Are we studying how to create a more resilient economy, structured so that it can cope with such shocks? Examining models to see how the crisis economy might be stabilized to prevent market crashes and currency crunches? (The economists I've asked all say that they haven't heard of anyone working on the crisis economy problem.)

An Oklahoma main street.

Is anyone planning, perhaps, a subsidized conversion to drip irrigation or hothouse agriculture? Or making agriculture do more with less? Pigs are terribly inefficient. You only get back about 15 percent of the calories that you feed them. Chickens manage 25 percent, beef 20 percent.

Vegetarians entirely bypass the middleman with the three-fold markup.

We can get trapped by our metaphors. Inadequate ones such as "gradual warming" produce tunnel vision, with all the dangers of being blindsided. So let us be on the lookout for handy metaphors, ones brief enough to use for a question to the speaker or in a letter to the editor. For example, there is a common assumption, both by some scientists and by many commentators, that the future is some gradual extrapolation of recent history. So remind them of the stock market's rise during the 1990s, when many people became accustomed to annual compounding of more than 10 percent. The people that projected this into the future soon discovered that many stocks could also fall by 75 percent. "I used to be retired," a friend quipped. "Now I'm merely unemployed."

Surprises happen. You may not know what and when and where, but you nonetheless can make sure that the infrastructure emphasizes stability. Just as we retrofit our old bridges to survive earthquakes, we should redesign our agriculture and our economy to survive hard knocks from climate surprises, pandemics, and economic panics. So far as I can tell, this isn't on anyone's agenda.

The water table is lowering rapidly in many places, another tragedy of the commons. Crops remove CO_2 from the air. However, irrigation not only raises the humidity (a heat-trapping gas) but reduces the amount of sunlight reflected back out into space via making the surface darker. Water scarcity is going to be a major problem in many places as the planet warms and the winds rearrange. Given the high percentage of river flow that is used for such wasteful irrigation practices, conversion to drip irrigation and hothouses will surely be on the agenda. Cities without a reliable water supply will likely no longer tolerate the existing division of water resources—and have to pay higher food prices in consequence.

Doing something is, of course, expensive. Then too, everything is labeled expensive unless it's business as usual. We can no longer afford business as usual. But there will be great business opportunities for the countries that develop the expertise and the needed new technologies before others do.

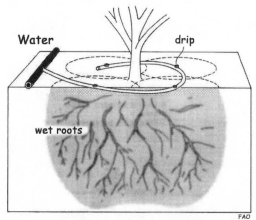

FAO

Drip irrigation method wets only the individual
plant and minimizes evaporation as well.

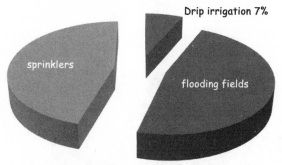

93% of U.S. irrigation uses wasteful methods

Agriculture now accounts for about 70 percent of world water use,
industry for about 22 percent, and towns and municipalities for 8
percent.

Center pivot irrigation (up to 35 percent of water is wasted via evaporation)

Spraying in this way has all of the delicacy of a fire hose.

The cookie-cutter crop circles in southwestern Kansas that result from using well water dispensed by a pivoting platform. The highway at right provides a sense of scale.

6

Why Deserts Expand

[T]he high Sierra meadows would likely die in the summer droughts. I love those high meadows, and the thought that I might be part of the last generation to see them, that the beautiful high Sierra might become like the blasted wastelands of Nevada, filled me with rage and grief.

—writer Kim Stanley Robinson, 2005 essay

Back before the 1998 El Niño fires in Borneo and the Amazon Basin, I assumed that a forest fire in a rain forest was improbable. Another early misconception was to think that because more equatorial evaporation will occur in global fever, more rain will fall and therefore—my mistake—the deserts should bloom.

Common sense failed me on both. In general, the big computer simulations of the physics show that the wet areas will become wetter, the dry areas will receive even less rainfall, and the deserts will expand into some heavily populated areas such as the Mediterranean, Cape Town, Perth, and southern California.

It takes some knowledge of the ups and downs of air to appreciate the reasons for this counterintuitive result. Some deserts are merely in a rain shadow of a mountain range—say, eastern Oregon—or just too far from an ocean—say, in western China or Mongolia—that there isn't much left for them. The world's major deserts, however, are mostly located between 20° and 35° from the equator for a different reason.

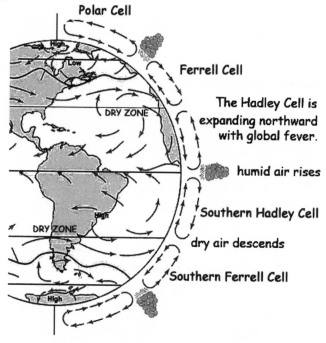

What goes up in all those thunderstorms in the tropics must come down somewhere. And, having lost most of the moisture on the way up, the air tends to come down very dry.

As the Hadley Cell expands
with global fever, so do the deserts.

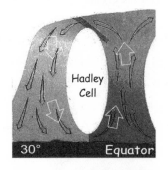

What goes up must come down somewhere. In the case of the warm, humid air that rises from the equatorial regions, it comes down dry about 30° from the equator, then turns back to make another loop. This is called the Hadley Cell after a London barrister, George Hadley, who in 1735 figured out the physics of how the return loop creates the trade winds.

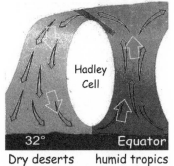

But now the Hadley Cell seems to be getting wider as a result of our global fever and its complications. That's pushing the dry zone farther away from the equator.

Dry deserts   humid tropics

This is called the Hadley Cell circulation and it's why we have the Sahara and Arabian deserts. In North America, we see the Sonoran desert at the same latitudes. (But why isn't Florida a desert? It's because that peninsula has the Gulf Stream on three sides to override the general tendency.)

There's another Hadley Cell to the south, resulting in the deserts in the Kalahari, Patagonia, and Australia. (There's also the Ferrell Cell between 30° and 60° and the

Polar Cell between 60° and the pole. The descending air at the poles makes them dry as well.)

The current global fever has the deserts on the march to higher latitudes. That's because the tropic's Hadley Cell is widening, changing where the air comes back down. Such climate creep spells big trouble for southern Europe and all around the Mediterranean Sea. The rain in Spain is mainly gone astray. Conditions as dry as the Dust Bowl will become the average climate in the American South-west, probably before midcentury.

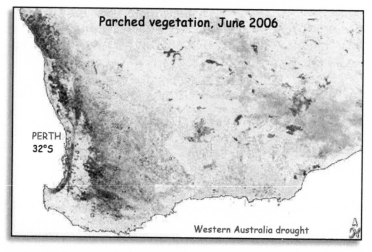

Australia is already having major problems with drought and encroaching desert. Perth had its average rainfall drop by 10 to 20 percent starting in 1976, resulting in stream flows being cut in half. Since 1997, the stream flows are down to a third of the 1911–74 average. In 2005, Perth's water experts rated the yearly chances of

catastrophic failure—no water coming out of the faucet—
at one in five.

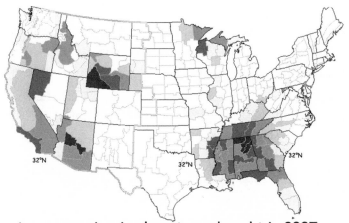

Areas experiencing long-term drought in 2007
www.cpc.ncep.noaa.gov/products/predictions/tools/edb/lbfinal.gif

It's been a similar problem in Cape Town which
presently, like Perth, just barely catches the westerly winds
that mark the end of the dry latitudes. The food and water
supplies of such major cities as San Diego at 32° from the
equator, Los Angeles at 34°, Cairo at 30°, Tel Aviv at 32°,
Cape Town at 34°, Perth at 32°, and Sydney at 34° are
particularly vulnerable to global fever.

Furthermore, the high places such as the Andes will
warm more than average, killing off the glaciers and much
of the water supply for cities such as Quito.

Both irrigation and drinking water will become a
problem in areas that rely on snowpack to store water for
summer use. With overheating, the same precipitation will
fall as rain, running off immediately. Even Seattle and San

Francisco will have problems from this, but southern California has had a serious water shortage for decades. Yet they keep building new houses in the desert without providing new sources of water—say, desalination plants run by carbon-free power. Perth just built a desalination plant to supply 15 percent of the city's water needs that is powered by a wind farm.

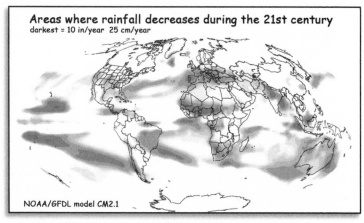

Areas where rainfall decreases during the 21st century
darkest = 10 in/year 25 cm/year

NOAA/GFDL model CM2.1

In the U.S., western and southern states get less rain. More serious decreases are seen in Central America, the mouth of the Amazon River, much of Brazil, all of southern Europe, the entire Sahel, southern Africa, northern India, Borneo, and most of Australia.

Besides making drought permanent in some areas during the coming decades, the Earth's fever is going to create deluges in other areas. Three straight days of heavy rain will become a more common event in areas that still have rain.

Recent evaporation helps to seed the next rainfall, as I mentioned earlier. That is going to be a big problem in the

Amazon as climate changes. Today, the flat bottom of the clouds (where the dew point is) isn't very high off the ground.

But with greenhouse warming, that flat bottom will move up higher in the sky, not mixing well with the recent evaporation. The clouds will continue westward until they run into the Andes. The rain they drop there will flow down the Amazon River as it does now, but the lush vegetation on the riverbanks will be gone—likely burned off during the onset of drought. It also means that many species of plants and animals will go extinct.

Morning clouds above the Amazon River in Peru.

It is estimated that burning off the Amazon and Asian rain forests will release the equivalent of twenty years of our usual carbon emissions—and that lacking their annual removal of CO_2 from the air, global overheating from other sources will increase by 50 percent.

We don't get big trees back very soon after a big fire, as the plant succession cycle has to build up to them starting with grasses and weeds, progressing to brush and trees. The vegetation reboots, but slowly—not recapturing much of the carbon that escaped into the air from combustion and decomposition until a century later. That could be too late.

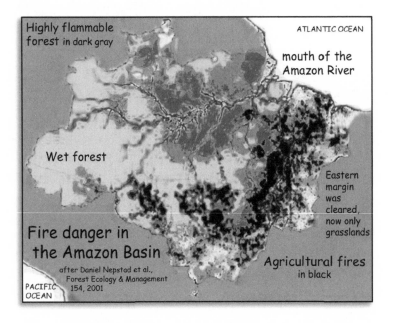

In the case of the Amazon Basin, it's even worse. Because of those peculiarities of the Amazon's water cycle, the climate models say that plant succession won't get past grass and brush. In the area on the southeastern rim of the

Amazon basin, already subjected to slash-and-burn forest clearing for a few years of marginal agriculture, the now-abandoned land has returned to savannah. The rain forest isn't coming back there.

Some "renewables" simply cannot be renewed within the relevant time frame. But this is a case of no new trees, period.

We must shockproof our food supply and our economy, and do this well in advance of encountering another regional or world-wide drought. To enjoy the long term, you have to survive the short term, over and over.

Too much focus on gradual trends, important as they can be in the long run, sets you up to be blindsided by the abrupt recurrence of a problem that you should have prepared for. Any notion of sustainability now needs to include surviving the flips.

Many people are still stuck in the outmoded balance-of-nature or ramping-up metaphors, not comprehending the true challenge of climate change. *Climate takes leaps and so we cannot merely back up if we overextend ourselves.* There are just too many aspects that are irreversible, such as the Amazon not re-growing its rain forest—or regaining its extinct species.

[The] amounts of carbon that may be going to the atmosphere
following Amazon droughts are probably big enough to accelerate
global warming. Currently trends suggest that a big chunk of the
Amazon forest will probably be displaced by fire-prone scrub
vegetation; global warming will probably exacerbate this trend.
                              —Amazon ecologist Daniel Nepstad, 2007

Our complex society relies on our being able to plant crops and
build cities, knowing that the rains will come and the cities will
not be flooded by incoming tides. When that certainty fails, as
when Hurricane Katrina hit New Orleans [in 2005], even the most
sophisticated society is brought to its knees.

But there is a growing fear among scientists that, thanks to man-
made climate change, we are about to return to a world of climatic
turbulence, where tipping points are constantly crossed.
                                    —science writer Fred Pearce, 2006

Fossil fuels helped us to fight wars of a horror never contemplated
before, but they also reduced the need for war. For the first time in
human history - indeed for the first time in biological history -
there was a surplus of available energy. We could survive without
having to fight someone for the resources we needed. Our
freedoms, our comforts, our prosperity are all the products of
fossil carbon, whose combustion creates the gas carbon dioxide,
which is primarily responsible for global warming.

Ours are the most fortunate generations that have ever lived.
Ours might also be the most fortunate generations that ever will.
We inhabit the brief historical interlude between ecological
constraint and ecological catastrophe.
                              —commentator George Monbiot, 2006

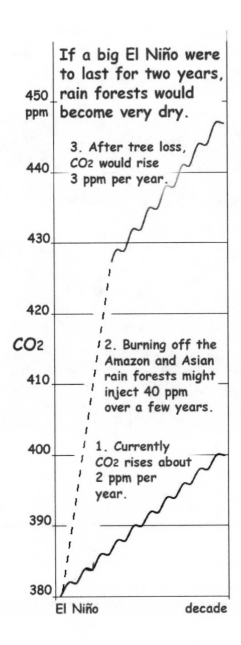

If a big El Niño were to last for two years, rain forests would become very dry.

3. After tree loss, CO_2 would rise 3 ppm per year.

2. Burning off the Amazon and Asian rain forests might inject 40 ppm over a few years.

1. Currently CO_2 rises about 2 ppm per year.

CO_2

450 ppm
440
430
420
410
400
390
380

El Niño   decade

The day after hurricane Katrina. Following the breach of two levees earlier in the day, the floodwaters rose in New Orleans on Tuesday, August 30, 2005.

Few people were killed during the storm itself; many died later because of an incompetent response by the government. (USCG photo)

7

From Creeps to Leaps

I'm in the dark as to how close to an edge or transition to a new ocean and climate regime we might be. But I know which way we are walking. We are walking toward the cliff.

—oceanographer Terry Joyce

The paleoclimate record shouts out to us that, far from being self-stabilizing, the Earth's climatic system is an ornery beast which overreacts even to small nudges.

—oceanographer Wally Broecker

So when does climate creep? Why does it sometimes take a flying leap?

It's hard to explain science without using a few metaphors. You need the metaphors to understand both science and history. And you need them to talk about the subject. Most of us manage to find simpler words when trying to explain our research at a cocktail party or coffeehouse conversation. Thanks to the social lubrication, we manage to sketch on napkins and create analogies to common objects and processes.

If the most knowledgeable people don't supply some metaphors, less applicable ones may be used by the journalists, ones that break down sooner. You need metaphors

to informally discuss the issues, whether in café convers-
ations or on talk radio. (As in asking questions after a talk,
these are all "short form" occasions during which you'll
likely be interrupted if taking more than a minute to make
your point. Practice your "elevator pitch.")

And if you hope to change the world, metaphors can be
politically powerful. Just recall the domino theory for the
containment of communist expansion, which served as a
major rationale for the Vietnam War. Everyone could easily imagine that row of dominos and its fate. Clearly, we
need some politically powerful metaphors for the fate of
our civilization, not to mention the rest of our Earth.

> [All] thinking is metaphorical, except mathematical thinking.
> What I am pointing out is that unless you are at home in the
> metaphor, unless you have had your proper poetical education in
> the metaphor, you are not safe anywhere. Because you are not at
> ease with figurative values: you don't know the metaphor in its
> strength and its weaknesses. You don't know how far you may
> expect to ride it and when it may break down with you. You are
> not safe in science; you are not safe in history.
>   All metaphor breaks down somewhere. That is the beauty of it.
> It is touch and go with the metaphor, and until you have lived
> with it long enough you don't know when it is going.
>   —poet Robert Frost

We all tend to assume that twice as much input yields
twice the output. This is, literally, "linear thinking." It
works for many things, such as paychecks for an hourly
wage. Time-and-a-half for overtime is a familiar non-
linearity. Climate is full of nonlinearities.

For example, consider windstorm damage to buildings. As the wind blows harder, more trees topple over. But at about 50 mph (80 km/h), things change. Just another 20 percent increase in wind speed to 60 mph and the insurance claims go up 650 percent. Overtime pay just doesn't take off like that.

More objects become airborne, striking buildings downwind and perhaps knocking something off. Effects snowball when detached objects stay airborne for long enough to hit another building and knock something loose from it. This cascade is why insurance companies are so alarmed at the prospects for stronger winds with climate change. That's why insurance premiums will rise far more than 20 percent—if insurance remains available at all. (And without insurance available, few will be able to get a mortgage.)

The most common way in which things "take off" is via exponential growth. That's when the size of the next increment is proportional to the present accumulation (the rich get richer).

Take compound interest: at 10 percent per year, your $100 increases to $110, then $121 after the second year, $133, $146, $160, $176, $194, and so doubles your money in the eighth year (rather than, without compounding, at the end of the tenth).

The same principle that compounds savings accounts also applies, at a much higher interest rate, to credit-card debt (the poor get poorer, and even more quickly).

Think of exponential growth as the Pothole Principle. That is, after all, how it impacts us almost every day. Teeth-chattering, tongue-biting impacts. Deferred maintenance, as when Seattle puts off fixing the potholes until next year, is a very expensive practice. It is not just two year's worth of wear and tear to fix the following year.

Genuine Seattle pothole

That's because the bigger the pothole is, the more rapidly it enlarges. When a tire can descend into the hole and hit its far edge more directly, it can break off a larger chunk of paving. And, perhaps, blow out the tire. The national highway in Madagascar has so many deep potholes that—out of consideration for their vulnerable tires—everyone drives on the muddy shoulders, just as they did before the road was paved.

Generally, when you see exponentially accelerating growth within an individual, you immediately think of cancer. Exponential growth in human populations also occurs. More babies mean more mothers a generation later, which means even more babies, etc. At its present growth rate, Kenya's population will double every eighteen years.

Once a leak starts enlarging, a dam can be rapidly torn apart. This is the Teton Dam in Idaho in 1976. The people living downriver had only an hour's warning and fourteen were killed.

An overflowing dam provides an especially dramatic example of the pothole principle. That's because water overflowing a dam eventually manages to find a crack and enlarge it. All of the overflowing water now tries to go through this little lowered section of the dam rim.

Eventually the water cuts a channel. The deeper the cut, the more water pressure is now behind the outflow. Thus water flows faster and carries away even more material.

Pretty soon large chunks are carried away by the thundering water flow, releasing the entire reservoir in a matter of hours. The same principle applies to leaks below the surface.

You do not want to live downstream of an earthen dam or dike. (As one expert said about the dikes in New Orleans, "There are only two kinds of levees: those that have already failed and those that will eventually fail.")

On some rivers, there are a series of dams that create a staircase of reservoirs descending toward sea level. If a surge arrives from an upstream dam collapse, it might damage the next dam in line via the overflow—a real-life domino effect.

Familiar with such exponential growth in a population, Rev. Thomas Malthus noted in 1798 that their resources were not likely to grow in the same way. A farmer's field may expand but not by an amount that takes into account how much it expanded last year.

This difference, he observed, would surely limit population numbers once the food supply was fully exploited. (A generation later, this Malthusian contrast between the two different growth curves in economics gave Charles Darwin an important clue that steered him to his theory of evolution by means of natural selection.)

This insight by Rev. Malthus shows why it is so important to ask, "Relative to what?" Things interact and so that's the first place to look. If they oppose, do they really balance out? All the time? Small fluctuations in strength, even if they don't last for long, can change everything. Consider a tug-of-war, where one team on the end of a rope tries to pull the other team across a line on the ground. Progress can be a gradual back and forth, the way that economists imagine supply and demand interact to determine the price of wheat. But the teams only have to be unequal in stamina and sure-footedness for a brief moment in order for a surge to occur.

There are many situations in climate where effects are delayed, as when warming speeds up decay in the soil or melts permafrost. Leads and lags can harbor surprises. So let me describe a common situation where everything seems to be "in balance"—but actually isn't.

Consider the increasingly booming business experienced by a new restaurant as the good word gets around. The Berkeley astronomer Richard Muller likes to explain why such a new restaurant can suddenly go bankrupt despite being packed with customers.

The important dynamic operating here is the traditional lag between receiving income and paying the bills (sound familiar?). "They get the income from the customers immediately, but they don't have to pay their bills until next month. As long as the business is growing, they're covered, even if they're losing money on every meal." The growing income this month means they always have enough to pay last month's smaller grocery bills.

"This goes on for a year," Muller explains, "until business stops growing, and suddenly they can't pay their bills." The restaurant hadn't been charging enough to cover its costs, and didn't realize it until too late. Flattening of growth became their tipping point.

Unreckoned environmental costs could make many aspects of our present economy look like Muller's new restaurant syndrome. Leads and lags give you a whole new perspective on many issues, such as why a flat economy is feared so much, why "growth" has become such a mantra.

Climate exhibits many leads and lags, some of which will change as global fever spikes. Most familiar is the seasonal lag in those locales where most of the rain falls in winter, while most of the plant growth occurs in the summer. Farmers come to rely on this separation and dread heavy rains in the summer. This flattens half-grown grain crops, which then rot.

The famines in Europe during the 1500s were due to exactly such "unseasonable weather," not the better-known cold winters of the Little Ice Age. Just imagine an unsettled, blustery late winter month (March in Seattle) becoming the summer standard. That's why the Irish shifted from planting grains to planting potatoes, a far less chancy crop because they hide in the ground rather than standing up straight.

They were a huge success, but the over reliance on them left Ireland vulnerable to a potato blight starting in 1848 that contributed to the famine deaths of over a million Irish. Not spreading your bets is a common beginner's mistake. Monoculture results from an excessive focus on short-term efficiency.

For decades, the climate problems have been framed as gradual creeps in temperature and CO_2. But climate comes from a web of interactions. Focusing on one of them at a time may miss the greater significance. Sometimes they oppose, sometimes one amplifies the effects of another.

The big setup for flips is when interacting processes have different characteristics. That sets the stage for "nonlinear" interactions that produce tipping points and flips. Say, one "team" is twice as sensitive to CO_2 or temperature as the other. Or one is nimble and the other is ponderous, as in some wrestling matches.

The atmosphere can mix up CO_2 contributions from Canada and Chile in only a year or two, while it may take a thousand years for ocean currents to mix waters from the Arctic and Antarctic Oceans. Yet the nimble atmosphere and the ponderous ocean interact, as when a warm pool heats the air above it. As the warmer air rises and pulls in the near-surface air from neighboring regions, winds are generated. And, vice versa, winds affect the ocean currents by pushing the waves. This helps cold deep waters to rise to the surface locally.

Sometimes we just dig ourselves in deeper. That's usually because of some combination of forces at work. A down-to-earth example is quicksand.

In my undergraduate days at Northwestern University, I had a job as the Saturday morning tour guide for the science and engineering building, showing visiting parents and prospective students around. There were two high points of this tour for most of the visitors. One was in the Department of Physics where, since I was a physics major, I turned on the model van de Graaff accelerator—while touching its dome. Thousands of volts soon made my hair stand on end (I had more hair then). The other was in the Department of Civil Engineering, where I turned on the water flow to convert a table of sand into a table of quicksand.

I had to use a water-filled doll as my stand-in. It sank, but only halfway—like water-filled humans do if we keep calm. I could make it sink further, however, by bubbling some air through the sand, making the water less buoyant.

That's why you should try to float, should you find yourself in quicksand, rather than flailing away and driving air into the sand beneath you in your struggle—converting it into quicksand.

Later I encountered this principle when taking flying lessons. Pilots are taught to avoid a situation known as getting caught on "the back side of the power curve." The normal part of the curve is what everyone knows about from driving a car: give it more gas, and you go faster. Airplanes have that additional feature where, at the slowest speeds when the nose is pointed up, increasing the power only makes the plane fly more slowly. To avoid a stall, where the plane isn't flying fast enough to support itself via lift on the wings, you can't just give it more gas.

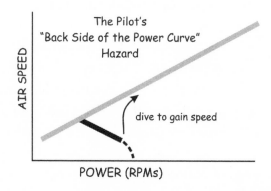

And how does the properly trained pilot get out of the situation? You gain airspeed the other way, by putting the nose down and diving, until enough speed builds up so that you pop across to the normal side of the curve, whereupon giving it more gas will indeed increase airspeed.

What if you don't have enough altitude to utilize the dive maneuver? (Say, you've just taken off and a stall threatens.) Lack of room to maneuver can be fatal. The same thing happens to whole societies.

Natural systems can also interact with agriculture to produce downside curves. One of the dangers of climate change is that we could find ourselves in situations where the harder we try to extract ourselves, the worse it gets. A familiar example is the intensification of agriculture that eats through the limited supply of topsoil.

The hillside plow turns rivers into mud.

Mule goes here

When drought makes some land unproductive, the remaining land is likely to deteriorate even faster for reasons of over-use and plowing high on hillsides (which may still have rain but, once plowed, the soil washes away in a few years). It takes many thousands of years to make new soil by weathering rock and, with no living tree roots to enlarge cracks in the underlying rocks, it takes even longer to replace what washed out to sea.

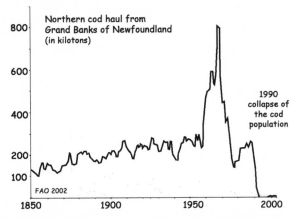

Northern cod haul from Grand Banks of Newfoundland (in kilotons)

1990 collapse of the cod population

FAO 2002

Overfishing provides another example of crashing from a lack of room to maneuver. By 1988, scientists said that the Grand Banks fish populations were on the brink of collapse and that, according to their models, the allowed catch must be cut in half.

Under pressure from fishing communities, the Canadian government only reduced the catch by 10 percent. By 1992, the collapse was complete. Then they halted fishing. "Compromising" with scientific warnings has a sad history.

> Laypeople frequently assume that in a political dispute the truth must lie somewhere in the middle, and they are often right. In a scientific dispute, though, such an assumption is usually wrong.
> —ecologist Paul Ehrlich

> Scientists know how to limit further damage, but governments know that the necessary changes could cause job losses and declines in tax revenues. Although a democracy needs sound public knowledge to help enlighten political actions, the public is spectacularly ignorant about many large-scale scientific issues.
> —author James Martin, 2007

That we don't know when a flood will happen has not prevented us from building dikes and dams. We now have building standards that help limit damage from the next randomly timed earthquake. At a minimum, business-as-usual prudence suggests that we must shore up our society's infrastructure for abrupt shifts in climate. No government seems to be doing this yet, even for the more common climate problems such as drought.

We can get trapped by our framing and our metaphors. Inadequate ones such as "gradual warming" produce tunnel vision, with all the dangers of being blindsided. One reason that "global climate change" is promoted as a replacement phrase for "greenhouse warming" is to broaden the agenda, to keep people from getting trapped by the familiar framework of "How warm it might be tomorrow" and reflecting on the unreliability of weekend weather forecasts. Greenhouse warming is simply too one-dimensional a concept, given all the droughts, dust, and high winds that are likely to accompany it.

However, the climate deniers' noise machine then started using "climate change" to mean "climate cycles," to take the focus off climate trends like the warming. That's why I like "Global Fever and its complications."

So "creeps" can take off by turning into exponentially growing "potholes." Collapse lurks. Unrecognized lags and leads can crash the system if climate alters them. You can dig yourself in deeper.

So what's a tipping point? Let us say you are pushing a baby carriage up a hill in the park. You can anticipate the effort needed to get up the next stretch of the path. But upon reaching the top of the hill, things abruptly change and you have to run after the carriage as it starts downhill.

If you must experiment with tipping points, just slowly lift up one side of a table. At some point, the dishes will begin to slide downhill. Lift even higher and, at some height, the table will flip over on its side. So even with mere furniture, there are two separate tipping points, a slip threshold and a flip threshold.

In the climate system, tipping points may be invisible until encountered. That's why studying ancient climate is so important. It shows us many of the past episodes of tip, slip, and flip.

We're operating this planet like a business in liquidation.
—Al Gore, 2006

We're altering the environment far faster than we can possibly predict the consequences.
—climate scientist Stephen H. Schneider, 2007

A cogent case has been made that one should pay more attention to low-risk but potentially catastrophic events, as opposed to the current focus on the "most probable" case. Those who would sneer that such an application of the "precautionary principle" would lead to paralysis are relying on an extreme caricature of the principle that has little resemblance to the way it is used in practice.

For example, if one is thinking about driving down a mountain road at night and has faulty headlights, knows that the ravine ahead has a rickety bridge over it, and has heard that there has been a storm that may have washed the bridge away, one would be quite justified in driving slowly or perhaps even postponing the trip, even if it was not known for certain that the bridge had been swept away. No doubt, those who disdain the "precautionary principle" would be quite happy to load their whole family in the car and put the pedal to the floor.
—climate scientist Ray Pierrehumbert, 2006

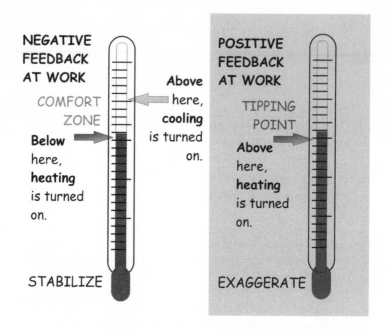

NEGATIVE
FEEDBACK
AT WORK

COMFORT
ZONE

Below
here,
heating
is turned
on.

Above
here,
cooling
is turned
on.

STABILIZE

POSITIVE
FEEDBACK
AT WORK

TIPPING
POINT

Above
here,
heating
is turned
on.

EXAGGERATE

8

What Makes a Cycle Vicious?

For droughts, it was just one damn thing after another—
not identical dominos falling into one another, but rather a
varied knock-on chain of causation. Rube Goldberg is said
to be the patron saint of biology because life forms feature
such long chains of causation. They have some virtues, as
they provide multiple ways to interfere with the end
product—medicines often work by interrupting the chain.
Perhaps the knock-on chains in climate mechanisms will
similarly offer some promising targets for intervention.

But there are also vicious cycles in which one action
repeats, building up steam. You'll need to understand
vicious cycles to appreciate the greatest climate change of
all time, "Snowball Earth," in which even the oceans froze
down into the depths (sea ice today is seldom more than
several meters deep).

Most people associate the term "feedback" with
suggestion boxes and critiques, with positive feedback
being good and negative feedback being bad. There are no
such connotations to the scientific use of the terms.
Negative feedback is commonly used to stabilize some-

thing. Let the room temperature fall below the desired temperature, and the modern thermostat turns on the furnace. If temperature rises too much, the cooling mode starts up instead.

One of the earliest industrial applications of negative feedback was the speed governor of James Watt's steam engine. The faster it spun, the more the spinning weights "stood up" and slowed the steam supply. Let the speed fall and the balls drop, the air supply opens up, and the speed drop starts to reverse. This enables the engine to run at a steady speed (to set it for a lower speed, just tighten up the linkage rod).

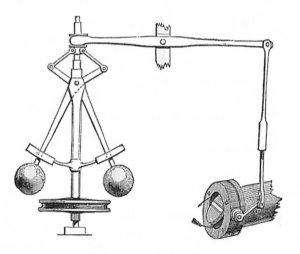

Negative feedback serves to hold speed constant.

All this is pretty obvious when you look at it working in a science museum, but analyzing it mathematically confounded one of the great physicists of the nineteenth century, James Clerk Maxwell. "Try as I may, its analysis defies me," he is reported to have said. There is no standard Cartesian cause-and-effect way of thinking about it; the logic becomes circular. The whole has become something more than the sum of its parts.

The real world is full of such negative feedback mechanisms where cause and effect get all mixed up. When I was a graduate student in the early 1960s, researchers were busy studying the feedback loops for adjusting our body temperature, our blood pressure, and our body fat. I can well remember how confusing it was to make the transition from simple cause-and-effect reasoning to the thinking needed for dynamic systems with feedback.

The Earth's wide swings in climate during the ice ages were greatly exaggerated by feedback. So too is our present global fever: it's those feedback loops that make a mountain out of a molehill. We'd better understand them thoroughly.

Most vicious cycles involve positive feedback. Besides compound interest and potholes, you have also likely encountered the squeal when a microphone picks up a sound from the loudspeaker and amplifies it even more. And more. Fortunately, the amplifier's power supply has a limit beyond which it cannot go, or the exponential growth of the sound wave might shake apart the entire universe.

There is a version of positive feedback for snow and ice on the timescale of days. When it is cold enough to snow, it can get colder automatically. That's because white surfaces reflect a lot of sunlight back out into space—about 90 percent of the incoming energy, as long as the snow is nice and white. Uncovered, the ground might have absorbed most of the incoming sunlight, changing it into heat. With snow, this usual input gets rejected, bounced out of this world. The same thing happens with sea ice capping an ocean. And so the world cools a little. The climate changes over the years.

With it cooler, snow covers even more of the Earth, and for more of the year. That's the vicious cycle. Colder means whiter, which means even colder.

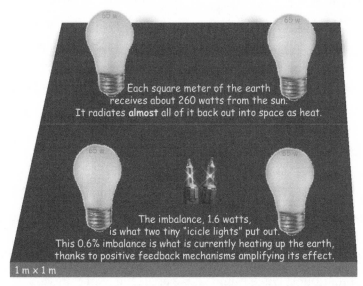

Each square meter of the earth receives about 260 watts from the sun. It radiates **almost** all of it back out into space as heat.

The imbalance, 1.6 watts, is what two tiny "icicle lights" put out. This 0.6% imbalance is what is currently heating up the earth, thanks to positive feedback mechanisms amplifying its effect.

1 m x 1 m

The reason that a small 0.6 percent imbalance can cause such big climate changes is that it is amplified greatly by a series of feedback loops, as when green replaces white in the Arctic.

You can see how the earth might become one big white ball rotating through space, with the remaining oceans frozen solid. Indeed, the earth might get stuck in that unhappy state, with no way to warm itself back up. This is known as the White Earth Catastrophe. Like a black hole, once you're in it, you can't get out. (It must not have happened, the reasoning went, because here we are, not frozen.)

This was one of those homework exercises for students until recently, when it was discovered that it had actually happened. Not once, but at least three times, all back in Precambrian times perhaps 700 million years ago. Something reversed each episode of Snowball Earth. It's called the greenhouse effect.

More CO_2 warms things up because it traps some of the heat that would otherwise escape into space. But a hotter earth also radiates more heat back out into space. This eventually brings solar input and earth's heat output back into balance, whereupon things settle down at some higher temperature. So it's as if more CO_2 nudged up the thermostat setting for the Earth. It's much like tweaking the linkage rod on that engine governor of Watt's.

During Snowball Earth conditions, parts of the ocean remained unfrozen because they were adjacent to hot springs—and that's how life forms survived. Here and there, a volcano poked through the snow and ice. A volcano emits a lot of CO_2. And the major route for removing CO_2 was missing. Normally some CO_2 is carried down by rain and reacts chemically with rocks. With no rain to "weather" the rocks, the CO_2 hung around in the atmosphere, accumulating.

Eventually the surface of the equatorial snow began to melt a little from the combination of greenhouse warming and summer sunshine. This started another positive feedback cycle that repeatedly warmed the Earth. What was it?

> Palaeoclimate data show that the Earth's climate is remarkably sensitive to global forcings. Positive feedbacks predominate. This allows the entire planet to be whipsawed between climate states.
> One feedback, the 'albedo flip' property of ice/water, provides a powerful trigger mechanism. A climate forcing that 'flips' the albedo of a sufficient portion of an ice sheet can spark a cataclysm. Inertia of ice sheet and ocean provides only moderate delay to ice sheet disintegration and a burst of added global warming.
> Recent greenhouse gas (GHG) emissions place the Earth perilously close to dramatic climate change that could run out of our control, with great dangers for humans and other creatures.
> —James Hansen, Makiko Sato, Pushker Kharecha, Gary Russell, David Lea, and Mark Siddall, 2007

Well, it's not surprising that you didn't know, as it's only recently that I've heard the earth scientists emphasizing it. (They call it an albedo flip.) But it is simple enough.

Fresh snow and ice reflect about 90 percent of the sunlight, with 10 percent remaining to heat up things on Earth. That's when skiers and climbers need really dark sunglasses on a clear day, with sun block to match.

With a little moisture forming on the snow surface during the day (what skiers near Seattle usually experience), the snow is no longer as bright and ordinary sunglasses suffice. Wet snow is slightly gray, reflecting only 70 percent of sunlight. This may triple the amount of solar energy retained by the snow pack. And so, unless there is a fresh snowfall to restore things, the surface warms up.

In Snowball Earth, the greenhouse warmth finally wetted the surface, which in turn caused even more snow to turn gray around the edges, and so it warmed some more. Pretty soon, the ice melted in the tropics, leaving a blue ocean that retained most of the solar energy that reached it. Eventually, when the ice retreated from the subtropics, positive feedback finally shifted into high gear and finished the melting job.

Saved by the greenhouse. But there is one little problem with going this route, one we wouldn't want to suffer through. It stays really hot and humid for millions of years after the ice is gone.

Getting rid of the greenhouse excess can be a slow business. Some of the CO_2 can be gradually stashed in the ocean depths. Photosynthesis cleans out some more, converting CO_2 into more plants. But the major way of removing CO_2 from the air, certainly back in the Precambrian era, is through the "weathering" of rocks. The CO_2 in the air combines with the water vapor to form a weak acid. When that rains on rocks, it is corrosive. Thus some CO_2 fails to return to the air. Instead the carbon compounds are washed down the rivers and sink in the ocean depths.

Weathering is so slow that, for many millions of years after the melt-off, the Earth remained quite hot, about $50°C/120°F$. That compares to about $15°C/60°F$ for a present-day global average temperature. Our current excess greenhouse could take a long time to counter, given that we've already cut down most of the forests.

The white areas of Greenland on this map are where the air temperature remained low enough in 2005 so that summer sunshine never succeeded in melting the surface layers, not even for a day. All of Antarctica is that cold too, except for the Peninsula. The light gray shows the areas that melted in 1992; the darker gray the additional areas that melted in 2005. In northwest Greenland, little melted in 1992, but more than a million square kilometers melted in 2005. Huff and Steffen (2005).

"Climate feedbacks" is a phrase you see in the news, especially when a newspaper editor has decided that the issue is too complicated for ordinary mortals to understand. But it really isn't so difficult, as you see in these two examples of positive feedback in climate change. Brighter-is-cooler-is-even-brighter in one direction, then green-house warming makes it possible to melt it all when getting progressively gray.

There is, of course, a normal heat-trapping effect, as those nineteenth century physicists realized. Things would be pretty cold if there were not. The "greenhouse prob-lem" is simply about the unfortunate change toward too much of a good thing. Everything is a poison at a high enough dose, even water. Now we know that we have to treat fossil fuels that way too.

What you may not have heard is that the most common heat-trapping gas is not CO_2 but humidity. There is a lot of it around clouds. That's the reason why those cloudless nights are cooler. Water vapor accounts for 60 percent of the normal greenhouse effect. In second place is CO_2, and third place goes to methane.

You hear about CO_2 effects on climate change because it is clear that cutting down forests and burning fossil fuels are things that are likely to increase CO_2 levels (so is the Portland cement-making process). But don't get distracted by the blame game, those arguments over whose fault it is. Even if the whole overheating problem were to turn out to be a brighter sun or methane seeps, we'd still have to do something to save our civilization, and CO_2 is the heat-trapping gas over which we have the most voluntary control.

Now note that humidity amplifies the CO_2 warming effect by about 50 percent. When CO_2 increases the storage of heat in the lower atmosphere, it promotes more evaporation from the tropical oceans. And thus even more warming. Positive feedback strikes, once again. A 1°C warming increases both humidity and rainfall by 7 percent. (The models had predicted that rainfall should only increase by 2 percent—but the data say otherwise.)

Besides amplification, has humidity caused a greenhouse by itself in the past? (Say, via changes in atmospheric circulation that allow more water vapor to circulate?) It is difficult to measure anything about average humidity from the usual paleoclimate records until you get to the time span of tree rings, but the geophysicists have been very clever about inventing new methods and perhaps some information will emerge.

What is, however, already clear is that the lower atmosphere (containing most of the humidity and clouds) has gotten thicker over the last half century, as well as warmer. That's more troposphere and less stratosphere. Now there is room for more water vapor in the atmosphere—and so more blankets are wrapped around an already feverish Earth as part of the reaction to the added CO_2.

Positive feedback is also found in systems that bounce the sunlight back into space before it heats up the Earth very much. The sunlight rejection mechanism involves things that make the earth lighter and darker.

For example, for about 9,000 years out of every 25,000 years or so, there is enough hot-summer monsoon rainfall to allow the Sahara and the Arabian Peninsula to grow grass. With the earth's surface darker, less sunlight was bounced back into space. Which, of course, helped warm things a little more.

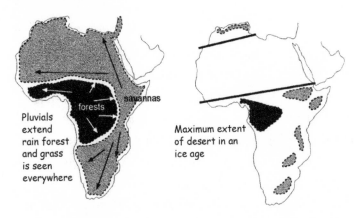

At the maximum extent of African savanna, about 6,000 years ago, most of Africa was green. During an ice age when even grass has a hard time, Africa is bright, reflecting much sunlight back out into space.

The last such "pluvial" period ended suddenly about 5,500 years ago. We know this as the period around 3500 B.C., when the great civilizations got started in the Nile and Tigris-Euphrates river valleys. In Egypt, a famous historical relief from about 3100 B.C., the mace head of Scorpion King, shows one of the last predynastic kings ceremonially cutting a ditch in a grid network.

Climate change can, it seems, stimulate innovation in government—at least back then. Tax collectors found that they needed to keep records of who paid what, and so writing was invented about 3200 B.C.

Melting ice means more of the dark ocean is exposed, allowing it
to absorb more of the sun's energy, further increasing air
temperatures, ocean temperatures, and ice melt. It seems that this
feedback, which is a major reason for the pronounced effects of
greenhouse warming in the Arctic, is really starting to kick in.
—climate scientist Ted Scambos, 2006

Changes in the land surface may be less dramatic than the melting of sea ice, but it is estimated that about 40 percent of the Earth's fever comes from cutting down trees over the centuries. When there is less uptake of CO_2 for growing trees, things warm even without fossil fuels.

But encouraging forests may not reverse this problem, as trees may also change the balance between reflection and absorption of sunlight. For urban trees with roads beneath them, more trees will mean less CO_2. But the present warming in high latitudes is causing the trees to grow farther north. Like melting the sea ice, this warms up the locality by absorbing more sunlight. Similarly, irrigation of sandy soils changes a light-colored surface into a dark one.

This isn't a major consideration in the tropics, but replanting there can be thwarted by the loss of biomass. Forest clearing for agriculture burns off the biomass that would have ordinarily gone into rebuilding the soil when the vegetation decomposed in the ordinary chain of events. And in the tropical rain forests, the web of roots is so efficient at recycling that the soil is quite thin in many places. Burn off the trees and you kill off the root system, allowing rains to wash away soil. Slash-and-burn agriculture exhausts the soil in only a few years, and so the farmers slash and burn somewhere else in order to feed their families.

They badly need a different agricultural economy. Indeed, given the global drought forecast, we all do.

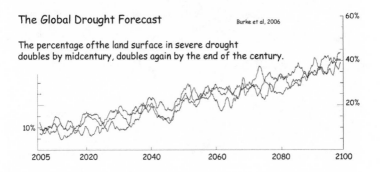

The Global Drought Forecast   Burke et al, 2006

The percentage of the land surface in severe drought doubles by midcentury, doubles again by the end of the century.

While climate change helped get civilization started, it is equally clear that climate change can collapse a civilization.

It depends, as Diamond makes clear in *Collapse*, on how people respond to the challenge. And how early they act (it's that weakening frog in the hot tub again).

Dust blowing off of the Sahara out over the Atlantic Ocean reflects sunlight back out into space. Also, the iron in the dust serves to fertilize the plankton in the surface waters of the ocean, thus taking more CO_2 out of the air and producing more oxygen.

9

That Pale Blue Sky

One surprise in recent years has been the effect of all of our modern pollution on cooling things. It has long been known that aerosols—sulfates and ash from power plants, smoke from fires, also dust and even sea salt—would reflect sunlight. Some light scatters down, making the sky a pale blue and the sunsets redder.

Some aerosols such as soot also absorb light. That means they warm the atmosphere while shading the surface. When you warm the atmosphere, you reduce cloudiness, allowing more sunlight through to warm the surface.

Until recently, no one realized the size of the cooling effect for aerosols from fossil fuels, what is now called global dimming. They turn out to have masked about a third of the twentieth century greenhouse warming—meaning that the CO_2 situation is even more serious than we thought a few years ago.

About two-thirds of the sunlight reaching the Earth is retained as heat. The rest bounces back out into space. Sand reflects a lot (it has "high albedo" in science-speak) but as rainfall improves and things turn into green grass and dark forests, less and less is bounced out. The tropical ocean bounces only about 10 percent. This means that when sea ice melts in the summer, you go from bouncing most of the sun's rays to absorbing most of them. You can see something similar when a blacktop driveway is covered by snow and then a small patch is cleared. The uncovered black patch heats up and melts the adjacent snow, soon clearing the entire driveway if it doesn't snow again and temperature remains near freezing.

Convert a desert into a suburb, and all of those dark parking lots and roofs now convert sunlight into heat (the urban "heat island" effect has repeatedly been trotted out by the disinformation propagandists to claim that global warning was a measurement artifact, even though it has long been clear that tundra and oceans have warmed up too).

When you irrigate the desert, you are incidentally warming the rest of the Earth in two ways: increasing the heat absorbed with the darker surface, and making the air more humid (and so a better greenhouse blanket).

Bouncing so much sunlight back out into space via aerosols ought to have cooled the Earth considerably, just as it does for several years after a major volcanic eruption. That it didn't cool shows that the greenhouse warming has been even larger than we'd realized—that particle pollut-ion has been cancelling part of the warming from the invisible pollution, CO2.

Economic problems can shut down many of the power plants (as happened in Russia in the 1990s). Remember leads and lags: the particles wash out of the atmosphere in weeks but the CO2 is only removed on a timescale of centuries. The world could suffer a heat wave following a more widespread economic crash.

Unfortunately, the pollution is also busy changing regional climates and it is feared that it may trigger major droughts in Asia and Africa, even burn off the Amazon. None of the climate mechanisms that we identify turn out to be as one-dimensional or cause-and-effect as we initially suppose.

COOLING side (changes since 1850)
-1.4 bright aerosols (ash, sulfur, etc.)
-1.0 more clouds
-0.2 cutting down dark forests

OUT OF BALANCE
by 1.6 watts/m²

WARMING side
+1.7 CO2 carbon dioxide
+0.5 methane (natural gas)
+0.3 CFCs
+0.3 ozone
+0.2 nitrous oxide
+0.1 soot (black carbon)
+0.1 sun brightening
+0.01 airplane contrails

When the most knowledgeable people about a subject start getting worried that big changes are going to happen, I tentatively side with them, not the neighbor who—emboldened by an ExxonMobil disinformation campaign aimed at providing sound bites for the coffee break—denies it all, despite knowing little about the subject.

I'm sure that it is only a matter of time before a disinformation campaign starts trying to sell you on the notion that pollution is good for you, that it is countering the greenhouse warming.

Just don't inhale.

> Supposing a currently envisioned low probability but high consequence outcome really started to unfold in the decades ahead (for example, 5°C warming in this century) which I would characterize as having potential catastrophic implications for ecosystems . . . Under such a scenario, we would simply have to practice geo-engineering . . .
>   —climate scientist Stephen Schneider, 1996

There are some schemes for managing the sunlight reaching the Earth, though none can be a complete solution. There is a clever proposal for placing a satellite at the L1 point, where the earth's gravity balances out the sun's gravity and so a satellite might stay put with minimal corrections. Attach a big sunshade to it and you might be able to diminish the sunshine reaching the earth by a few percent.

A high haze around the earth that enhances the amount of sunlight that gets bounced out is another way of

tweaking the earth's energy balance. It's even "natural."
The Mount Pinatubo eruption in June 1991, cooled the
earth's surface by about 0.5°C during the following year
with all the sulfur compounds that it injected into the
upper atmosphere.

Mount Pinatubo in 1991

Paul Crutzen, the atmospheric chemist who won the
Nobel Prize for his analysis of the ozone hole created by
refrigerator gases, has considered the use of sulfur to make
the upper atmosphere reflect some sunlight back into
space and thus cool the planet in an emergency.

The chemistry for the stratospheric-sulfur strategy is
about the same as for the sulfuric acid generated by
burning coal which causes acid rain. It currently leads to a
half million premature deaths every year. But the
stratospheric-sulfur strategy is not, as some newspapers
have proclaimed, going to bring back acid rain. (The same

mistake was made about ozone: it's bad down here but good high up.)

His proposal involves bypassing the weather of the lower atmosphere and using balloons or airplanes to distribute the sulfur in the stratosphere. It would stay there for several years, rather than washing out in a week as it does from coal plant emissions. So the amount needed for enhancing global dimming is only a few percent of the sulfur already being added by burning coal. But there's nothing special about using sulfur; a white powder such as diatomaceous earth would also serve.

The Earth's brightness
from NASA's 2002 Blue Marble Project

If the planet's [reflection back into space] dropped by just a tenth from today's [30 percent] level, to 27 percent, the effect would be comparable to a fivefold increase in atmospheric concentrations of carbon dioxide.

—atmospheric scientist Veerabhadran Ramanathan

The problems with all such schemes is that they are partial, not addressing the other effects of excess CO_2. If you thought that acid rain was bad for lakes and ponds, just wait until you see (in chapter 14) what happens to the oceans when the CO_2 concentration in the sea water rises. Acidified ocean is likely to ruin the tiny plankton that sink carbon to the ocean floor, the planet's major mechanism for sinking excess atmospheric carbon for million-year periods.

Still, the engineered high haze should be useful as a tweak as we try to back out of bright pollution without making heat waves worse. Think of such strategies as like the attitude-control thrusters on a spacecraft, which are not suitable for general propulsion but do allow for small-scale tweaks.

The discussions of geo-engineering heard in the science policy community have an air of unreality about them, as they assume some sort of global scientific consensus before deploying anything. I think it more likely that political leaders under pressure to "do something" because of a persistent heat wave will order their power plants to produce more aerosols to achieve local sunscreen, then order planes flying over their territory to use high-sulfur jet fuel, and later order their air force to disperse sulfur in the stratosphere—all without international consensus or scientific wisdom. Done in that fashion, they will probably make things worse.

Our dirty-power situation reminds me of that dilemma that pilots face when a stall threatens at a time they aren't very far off the ground—and lack maneuvering room to regain airspeed.

Experts are now saying that we only have a decade to get carbon emissions under control before we start getting into the zone of triggering major droughts and more rapid rise in sea level farther down the line.

We too could lose our maneuvering room and crash.

Had humans found bromine cheaper or more convenient to use than chlorine [for refrigerator gases], it is quite likely that by the time Paul Crutzen and his colleagues made their discovery [in 1974], we all would have been enduring unprecedented rates of cancer, blindness, and a thousand other ailments; that our food supply would have collapsed; and that our civilization itself would been under intolerable stress.

And we would have had no idea of the cause until it was too late to act.

—biologist Tim Flannery, 2006

Farmers often "clear the land" by setting
fires before planting. Satellite monitor-
ing of fires shows how intense the
practice has become. Even if the CO_2 is
soon reabsorbed by the plant growth,
that's a lot of soot constantly circulating
to absorb sunlight and warm the air. It
also modifies precipitation downwind.

Kerosene torch used for
setting agricultural fires

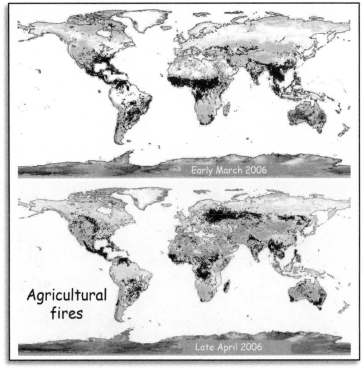

Early March 2006

Agricultural
fires

Late April 2006

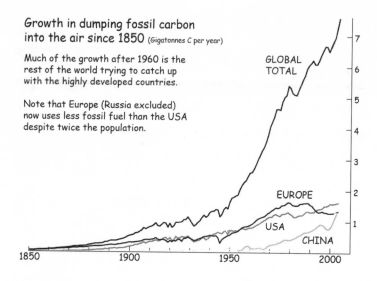

Growth in dumping fossil carbon into the air since 1850 (Gigatonnes C per year)

Much of the growth after 1960 is the rest of the world trying to catch up with the highly developed countries.

Note that Europe (Russia excluded) now uses less fossil fuel than the USA despite twice the population.

GLOBAL TOTAL

EUROPE

USA

CHINA

1850   1900   1950   2000

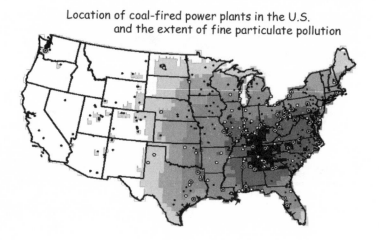

Location of coal-fired power plants in the U.S. and the extent of fine particulate pollution

10

Slip Locally, Crash Globally

Greenland is a different animal from what we thought it was just
a few years ago. We are still thinking it might take centuries to go,
but if things go wrong it could just be decades. Everything points
in one direction, and it's not a good direction.

—glaciologist Richard Alley

The common notion is that ice sheets build up, layer by
layer, when it's cold. And that the reverse process works
the same way, the sunshine melting the ice surface and
slowly eating down through the accumulation. And so sea
level slowly rises over the centuries.

Surface runoff and thermal expansion of the oceans is
what, for the most part, is used to estimate sea-level rise in
the 2007 report of the Intergovernmental Panel on Climate
Change (IPCC), the major international body that
evaluates the science for policymakers.

But this is seriously incomplete. It is merely the part of
the problem which can be computed. Unfortunately,
Greenland and Antarctica are big mountains of ice. Like a
pancake, the melting ice sheet spreads out. An even better
analogy is a scoop of ice cream fallen on the sidewalk. As it

melts, the height goes down and the sides expand. That's what was happening the last time that ice sheets covered up Seattle and plowed into New York's Central Park. But most of the present-day ice sheets are close to saltwater and, when ice reaches the shoreline, it is pushed into the water and instantly raises sea level—even without melting. The dynamics of iceberg production are very different from those of melting in place.

The speedups in Greenland outlet glaciers—doubling and redoubling before 2006—were primarily from the removal of obstructions to downhill flow. A glacier may plow into the ocean without floating, perhaps freezing to the bottom. But terminal melting occurs because the sea water is warmer. Further, the sea water starts digging farther under the glacier's snout, dissecting back to form an ice shelf. Tides now lift and drop the overhang, opening up a hinge of fractured ice—and so eventually another iceberg breaks off and sets sail. The warmer the sea water, the faster this occurs, and so the faster the rest of the ice can flow downhill. That's primarily what's been going on.

But if you fly over the Greenland ice cap (any summer nonstop between northern Europe and Vancouver, Seattle, or San Francisco will do, or you can use Google Earth), you will see a series of pretty blue lakes forming which are capable of speeding Greenland's collapse. Leading away from a nameless lake, you will see a nameless creek—that abruptly ends. That's because it found an express route to the bottom of the ice cap. Such a vertical channel, called a moulin, forms the world's tallest waterfall.

The water greases the skids, serving to float the ice sheet downhill to the ocean, where icebergs export the melting job to warmer places. When that happens, Greenland will not be merely melting, but collapsing. It will be like a runaway ice-cube machine.

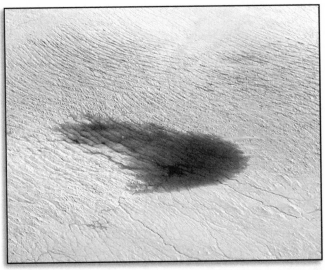

Greenland melt water lake, draining down a crevasse. Some large lakes have disappeared overnight. (Photo by WHOI oceanographer Sarah Das, 2003).

The ponds are located at the lower elevations where the temperature rises above freezing in the afternoons. And so the ponds and moulins are well positioned to affect the outlet glaciers that serve to buttress the central ice cap. Collapse the buttresses, and the rest can start flowing.

No amount of subsequent cooling can undo this structural damage to the base of the ice sheet. Stopping the melt will not stop the collapse and the iceberg production, though it can slow it somewhat.

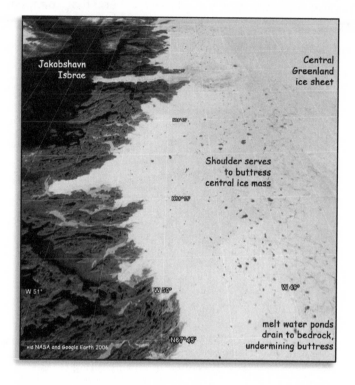

Looking north above the west coast of Greenland. Because the summers have gotten hot enough (this shows 2006), the shoulders of the central Greenland ice mass are now covered with pretty blue lakes of melt water. They drain to bedrock and some of the water may become trapped underneath, like a blister. Enough such blisters would set up a collapse that would quickly push a lot of ice off the island, raising sea level long before it actually melted.

Some of the water under the ice sheet refreezes. That's what traps the rest of the water and keeps it from later shooting back up the hole like a geyser. If the trapped water comes under pressure from glacial movement, it may not refreeze. That's because liquid water has to expand when transforming into a solid form. With all that weight of the ice sheet bearing down from above, the trapped water remains supercooled. Soon part of the ice sheet is resting on a liquid cushion (rather like a blister on your foot), floating above the rock whose surface irregularities, called pinning points, usually anchor the ice sheet to the bedrock. (This process hasn't yet been studied in Greenland's moulins, but there are good studies in Iceland where the trapped water was melted on the bottom farther uphill by volcanic hot spots.)

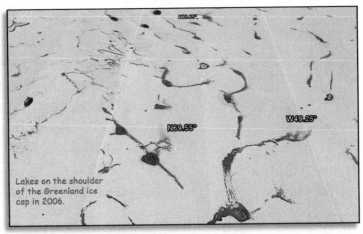

Lakes on the shoulder of the Greenland ice cap in 2006.

Close-up of the shoulder's melt water lakes, each with its own vertical drain to bedrock. Richard Alley notes that because water is 10 percent heavier than an equal volume of ice, "Water-filled cracks, more than a few tens of meters deep, can be opened easily by the pressure of water. Ponding of water at the ice surface increases the water pressure, wedging cracks open."

Normally, ice flows by deforming midlevel ice, with the foundation staying stuck on the pinning points, like the spreading batter on a waffle iron. Unstuck and lubricated, the base of the ice sheet can now spread out like ice cream on a dish.

And so the ice mountain surges out at the edges at speeds that can reach a mile a month. I once thought that glacial advance meant that the ice was building up. True enough when an ice age begins, but most of the glacial advances in the world today are from breakup's pancaking.

What if it now cools down and no more pretty blue ponds form on the surface of the ice sheet? Surely, you say, that supercooled water forming the cushion will become lost over the years and the ice will settle back down on bedrock. Maybe, but by that time, the base of the ice will have been destroyed in other ways. As ice moves over bumps, cracks open up and the water, under very high pressure, jets into the openings and cuts channels. The friction from this heats up the place, making more holes to jet into.

This process shatters the ice and turns the base into really rotten ice, the stuff that climbers hate for good reason. It is easily broken and nothing like the hard ice of the original base that was conveniently frozen to the irregular bedrock. Think of it as rotting away the foundations of the ice sheet.

So there is a legacy of the pretty blue ponds created by atmospheric warming. If the ice sheet is near a coastline, ice falls into the ocean in big chunks and sea level rises, a process which is much faster than Greenland melting in place.

Greenland's largest outlet glacier, draining 7 percent of the central ice sheet, is Jacobshavn Isbrae on the west coast. It's one of the galloping glaciers of Greenland, having doubled its speed between 1997 and 2003. The glacier itself is in the far background; what you see is the fjord, the icebergs, and the floating ice that is emerging into Baffin Bay. (Photograph by the glaciologist Koni Steffen.)

The water level in a glass of iced tea doesn't change when the ice melts. Ice shelves in Antarctica, when they break lose, get a lot of press. Then we see another round of talk-radio disinformation that suggests the scientists must be pretty dumb not to recognize that it doesn't change sea level by even an inch, ha-ha.

I used to be more understanding of this, reasoning that perhaps they didn't read far enough to realize that the big issue was making room for more ice to slide downhill, a plug having been removed. The crafters of disinformation always leave that out.

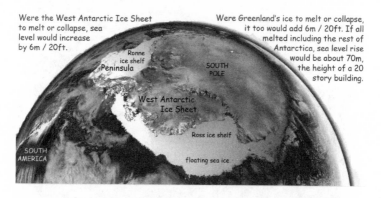

Were the West Antarctic Ice Sheet to melt or collapse, sea level would increase by 6m / 20ft.

Were Greenland's ice to melt or collapse, it too would add 6m / 20ft. If all melted including the rest of Antarctica, sea level rise would be about 70m, the height of a 20 story building.

Ronne ice shelf
Peninsula
SOUTH POLE
West Antarctic Ice Sheet
Ross ice shelf
SOUTH AMERICA
floating sea ice

Saltwater can get beneath ice shelves and unstick them from the bottom, exposing them to the tides every day, and eventually breaking them up. While this doesn't raise sea level directly, it is like uncorking a bottle. The ice streams uphill now meet less resistance and speed up. More of the ice sheet is pushed into the sea, instantly raising sea level. (Image adapted from NASA's "A Tour of the Cyrosphere.")

Collapse, not melt, is the operative concept. The sea level is rising, mostly from thermal expansion but increasingly from ice additions. How fast that happens is, I suspect, largely a function of ice sliding sideways, not melt rate itself.

In the past, how fast did sea level rise? Climate scientists now have good records of the ups and downs of sea level for many millions of years. The downs are due to giant ice sheets slowly building up in the higher latitudes. The ups can be much quicker, thanks to collapse. The sea level fluctuates by about 130 m during an ice age. That's the height of a forty-story building.

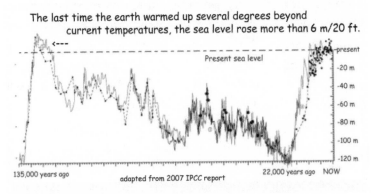

The last time the earth warmed up several degrees beyond current temperatures, the sea level rose more than 6 m/20 ft.

Present sea level

present
-20 m
-40 m
-60 m
-80 m
-100 m
-120 m

135,000 years ago   adapted from 2007 IPCC report   22,000 years ago   NOW

When a lot of water was taken out of circulation and piled up as ice sheets at high latitudes in Alaska, Canada, Greenland, Scandinavia, and Antarctica, the oceans weren't as full as today. That exposed a lot of new real estate, much of which turned green. The cooler oceans also grew more phytoplankton, the little photosynthesizing plants at the bottom of the food chain known as marine algae. The high winds added more dust to fertilize their growth. The productivity increase helped haul down the CO_2 by one-third—and that's about what we need to do today to reverse climate change.

Later, when things warmed up starting about 15,000 years ago, some of that ice melted and the sea rose to near its present level by about 7,000 years ago. That's how we got our present shorelines and that's when complex human societies got started.

This rewarming occurred because the summers were hotter in the higher latitudes. The earth's axis was then tilted more than it was this last time around, putting the sun higher in the sky when beaming down on Greenland. While the winters were also colder up there, this didn't balance out the hotter summers. Even when more snow falls in winter, the summer days get hot enough to melt the surface and turn it gray, enabling three times as much heat to be absorbed rather than being reflected back out into space.

As skiers know, if the daytime temperatures get up above freezing, even for a day, the subsequent skiing will be inferior until it snows again—but at least the glare isn't so bad. In an era of global fever, glare is good.

These [rapid Greenland ice] flows completely change our understanding of the dynamics of ice sheet destruction. We used to think that it would take 10,000 years for melting at the surface to penetrate down to the bottom of the ice sheet. But if you make a lake on the surface and a crack opens and the water goes down the crack, it doesn't take 10,000 years, it takes ten seconds. That huge lag time is completely eliminated.

The way water gets down to the base of glaciers is rather the way magma gets up to the surface in volcanoes—through cracks. Cracks change everything. Once a crack is created and filled, the flow enlarges it and the results can be explosive. Like volcanic eruptions. Or the disintegration of ice sheets.

—glaciologist Richard Alley

When things overheated several degrees about 125,000 years ago, much of Greenland melted and sea levels became at least 6 m higher than today, maybe 8 to 10 m judging from the coastal records from Brazil. That's probably an underestimate—I've seen estimates of 25 m for sea-level rise over the next century or two. It depends on how much the West Antarctic Ice Sheet contributes. It has two large ice shelves that could shatter much like Larsen B did in 2002. That would uncork the ice uphill.

But first let's take a look at what a mere 6 m rise in sea level will do to all of the people who live near the seashore. More than a third of the world's population now lives within commuting distance of the shoreline. Seven of the ten largest cities are situated on coastlines threatened directly by rising sea levels: Tokyo (34 million people, as many as all of California), New York, Sao Paulo, Mumbai (formerly Bombay), Los Angeles, Shanghai, and Jakarta (17 million). New Orleans and Miami are far down the list.

A lot of people live on river deltas, right on the front line of sea-level rise. The Netherlands, on the delta of the Rhine, has become the world's most densely inhabited country. Bangladesh is the fifth most populous country in the world with 144 million people, half living on the flood plain of the Ganges.

This analysis only addresses the creeping physical changes, not the associated societal leaps backward that result from resource wars and genocides. Climate change has been a major driver for wars and rebellions. In the pre-fossil-fuel-fiasco era, a cooling climate doubled their

numbers in eastern China. In the present era of rapid
warming, some of the same mechanisms will operate:
instead of a shortened growing season from cooling, we'll
see warming and drought eliminate that second crop per
year.

Local slips, especially in Greenland and West
Antarctica, have the potential to crash human populations.

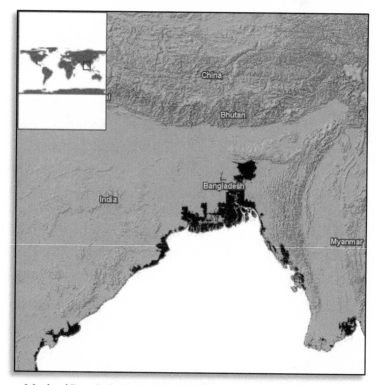

Much of Bangladesh, along with Calcutta in India, will be
permanently flooded by a 6 m rise in sea level. This will displace
seventy million people in a region already densely populated..

Bangladesh would lose 17.5 percent of its territory to a 1-m rise
in sea level, displacing at least 13 million people.

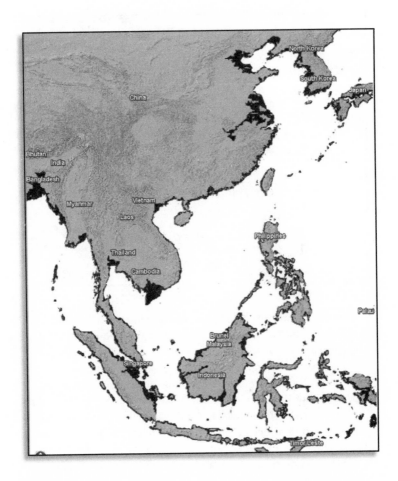

With an extra 6 m, the major river deltas of Asia are underwater.
They are heavily populated areas.

On China's river deltas, 72 million people would be threatened
with increased coastal flooding by a 1-m sea level rise. Egypt and
Vietnam, two other developing countries with large, unprotected
river deltas, each have eight to ten million people living within
1 m of high tide.

The entire North Sea coastline of Western Europe is severely challenged by a 6 m rise in sea level, as is Venice and parts of the English and French coastline.

The Dutch control flooding from the Rhine by dikes but storm surges arrive on the northern coastline. The surges from the North Sea storm of 1953 flooded extensive areas of the Netherlands and the Dutch have since built a barrier system like that seen on the River Thames, which keeps London from temporarily flooding. The light areas are population centers (a 6 m rise covers the dark areas; you can still see the population dots as gray areas within the dark areas).

Most of the sea-level maps were produced with the digital elevation mapping software at the University of Arizona, thanks to Jonathan Overpeck and Jeremy Weiss.

The Netherlands is the most densely populated country in the world today. Land reclamation is a centuries-old practice. The Dutch meticulously construct coastal defenses to a 10,000-year storm standard and carefully restrict development. They will likely keep adding to their sea-wall system. If the dikes were to fail, this is what a 6 m rise in sea level would do to low-lying areas of the Rhine delta.

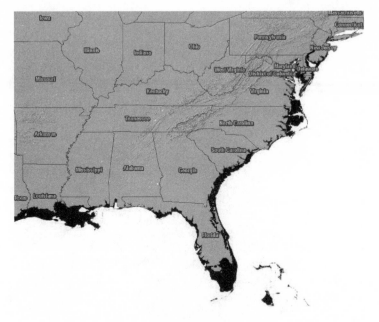

The source of a major refugee crisis for the U.S.: the Gulf Coast and Atlantic Coasts have many low-lying areas that will be flooded by a 6 m/20 ft rise in sea level. This is approximately the same territory that was underwater 125,000 years ago when the earth warmed several degrees above our present global average temperature—and that's equal to the most optimistic of the predictions for twenty-first-century fever.

Long before sea level reaches that height again, storm surges will reach much farther inland and destroy the homes of many millions. In Florida alone, 15 million people will be affected.

11

Come Hell and High Water

Indeed, its ever spiraling insurance bill resulting from severe
weather events and its growing water shortages in the west mean
that the United States is already paying dearly for its CO_2
emissions.

—biologist Tim Flannery, 2006

In terms of extreme weather events, it's worth noting that
the United States already has the most varied weather of
any country on Earth, with more intense and damaging
tornadoes, flash floods, intense thunderstorms, hurricanes,
and blizzards than anywhere else. With the intensity of
such events projected to increase as our planet warms, in
purely human terms the United States would seem to have
more to lose from climate change than any other large
nation.

Here I am going to address glacier-induced earthquakes
and the speed of sea-level rise, but the shorelines illustrat-
ed in this chapter will mostly be those of the United States.
I beg the indulgence of my international readers but, as

you know, my country is most people's favorite candidate for the big country which has cumulatively contributed the most to the problem (though currently China adds more each year).

Let's start with the minimal sea-level rise for the twenty-first century, which is about 0.5 m (20 inches). The inundation is about two-thirds of that we'd get from a 1.0 m rise, given the way elevations rise near U.S. shorelines. And since most glaciologists I know think that the IPCC figure is an underestimate, I'll briefly mention what a 1 m rise will do.

About the first meter of sea-level rise will submerge the Maldives. Mali, the country's capital, is already a walled city. The country lost 10 percent of its land in the 2004 Indian Ocean tsunami. For scale, that's a cruise ship just behind the island, to the right of center.

It's the first meter (3.3 ft) that will destroy Miami and most coastal areas in Florida. Furthermore, the damage will occur in episodes such as storm surges, long before average sea level rises 1 m.

Only 1 m (3.3 ft) of rise in sea level will inundate most of Miami and other coastal cities.

Before then, however, storm surges will have destroyed many populated places even farther inland.

USA: Florida

Weiss and Overpeck
The University of Arizona

The black areas are those which are above sea level now but will be underwater later.

The insurance people talk of a "hundred year flood" and whether a community is protected against it. Communities with insufficient protection for a hundred-year flood are labeled as being in a flood zone—and so they pay more for insurance.

More than half of the U.S. population currently lives in counties located along the 20,000 km (12,000 miles) of coastline. Major cities such as New Orleans, Tampa, Miami, Baltimore, Philadelphia, New York, Boston and Washington, DC, will have to upgrade flood defenses and drainage systems. But are we going to build 8,000 miles of seawall along the Gulf and East Coasts?

Storm surges that have occurred about every hundred years are going to become much more common, even for only a 1 m addition to sea level. For New York City, the hundred-year flood of the twentieth century will be experienced about every four years.

Now for the other end of the twenty-first century range of estimates, the 6 m / 20 ft of the last chapter. Most scientists who say 6 m is possible think that only 3 m of it could occur this century. But, since there has been a tradition of underestimates, I'll spell out the consequences of 6 m.

From home, I can walk down to the shoreline in a half hour—but I happen to live atop a hill over 100 m high, not likely to be flooded out. Yet like most Americans, I have relatives who live in places that will be ruined by a 6 m sea-level rise, such as out on the end of Cape Cod (the "arm" is going to be "amputated above the elbow" by rising sea level). Other relatives have a lovely new place on the Gulf of Mexico which will be totally underwater.

There's a sixth-floor family apartment in Florida which took a direct hit from three hurricanes in only two years. The bottom two floors of the building will be flooded. No more pool or parking lot. Initially I imagined looking down at a fleet of water taxis, maybe even gondolas a la Venice. Alas, the new shoreline would be miles away to the west and the building, exposed to the open ocean, would not last for very long before the waves pounded it apart. Perhaps we should put up posters in such areas, reminding people that where they stand was underwater the last time that the Earth ran a 3°F fever.

Southern Florida is entirely gone when sea level rises by 6 m. Orlando loses its eastern suburbs. Most of the 15 million people now in fast-growing Florida will eventually find their property uninhabitable and worthless.

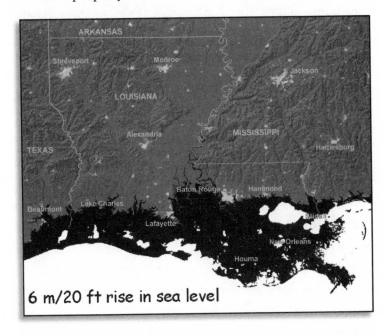

6 m/20 ft rise in sea level

Louisiana will also lose its most heavily populated real estate and New Orleans will be permanently flooded. Chesapeake Bay will become twice as wide. Washington DC, New York City, and Boston will lose substantial areas unless massive earthworks are constructed.

On page 133 is a simulated aerial view from Google Earth, looking west from Capitol Hill to the Washington Monument. The darkened overlay shows the inundation from a 6 m/20 ft rise in sea level, resulting from a $3^{\circ C}$ / $5^{\circ F}$ global fever. That's the White House at upper right.

Storm surges will attack both Capitol Hill and the White House. Permanently flooded are most of the museums along the Mall, most war memorials, the National Academy of Sciences, and all of the major government buildings along Pennsylvania Avenue. Perhaps they will name the ruined swath for one of the party-now-pay-later politicians of the early twenty-first century who didn't take seriously the responsibility to future generations.

One can, of course, imagine that the Army Corps of Engineers will build giant dikes to channel the higher river, much like the ones looming over New Orleans that contain the Mississippi River. But I find it difficult to imagine continuous sea walls on open coastline all along the Atlantic and Gulf Coasts. And turning limestone into cement releases its carbon into the air—so both precautions and storm damage are likely to create more greenhouse gases, meaning that burning fossil fuels has to be reduced that much further.

Then there's New York City. Manhattan will lose half its width below Midtown. Queens and Brooklyn will lose substantial territory, as will western Staten Island. Large areas of New Jersey will be underwater, including Newark and Jersey City. All four major airports will be underwater, as will Yankee Stadium. Protecting Long Island's coastline from Brooklyn eastward is going to be difficult and expensive, requiring both dikes and storm barriers.

Even at the present sea level, there is similar vulnerability during a hurricane. "One of the highest storm surges possible anywhere in the country is where Long Island juts out at nearly right angles to the New Jersey coast. They could get 25 to 30 ft [about 8 m] of storm surge," said the director of the U.S. National Hurricane Center in 2006.

Proposals to build three storm surge barriers (and surely a fourth to protect Jamaica Bay and JFK airport), like the Thames Barrier in London, have gone nowhere even in the city of hedge funds where they ought to know about hedging their bets. A hurricane that came ashore somewhat north of Atlantic City, New Jersey, would most efficiently drive a storm surge right up the Hudson River.

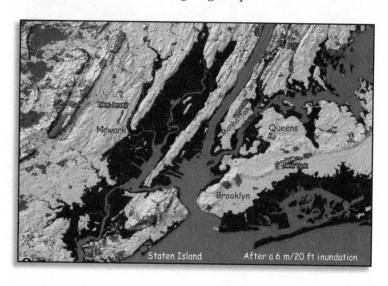

But it isn't just hurricanes we have to worry about. Severe winter storms that sit offshore farther south create strong winds, initially out of the northeast (hence the name nor'easters). They can produce large storm surges. Unlike a hurricane whose storm surge is gone a few hours later, the spiral center of a nor'easter may sit there for days without moving, continuing to pile up water inland.

Like London, New York City has a great deal of underground infrastructure that is vulnerable to saltwater flooding now, and the floods will occur much more frequently later this century.

During the December 1992 nor'easter, ocean water flooded the Hoboken train station, short-circuiting the entire New York City subway system. It took ten days before service was fully restored.

If there are rivers behind them, surge barriers can only be closed for a short time. For example, the Thames begins to back up when the Thames Barrier is closed. This means that a constant sea-level rise must be handled differently, as when dikes are built inland along river banks to allow for a higher river.

"If just one flood broke through the Thames Barrier today, it would cost about £30 billion in damage to London, roughly 2 percent of the current UK GDP," observed Sir David King, chief science advisor to the British government.

7 m
23 ft
sea level
rise

Lower Manhattan street level (World Trade Center, pre-2001)

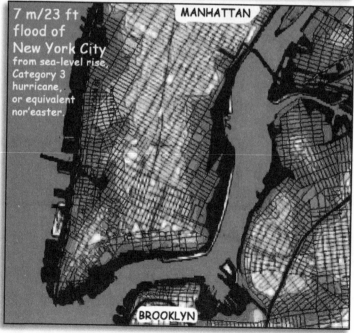

7 m/23 ft
flood of
New York City
from sea-level rise,
Category 3
hurricane,
or equivalent
nor'easter.

MANHATTAN

BROOKLYN

Storm surges will be the immediate "cause of death" of the island nations without elevated land. Global warming from our pollution is what sets up the new high tide line, but battering waves will scour such islands long before high tide reaches the island's high ground. It seems safe to predict that some climate change denialist will insist that the poor country was done in by a bad storm, not global warming.

To protect New York City and low-lying areas of New Jersey near the Hudson River, three Thames-like surge barriers have been proposed. A fourth could also protect Jamaica Bay and JFK airport, though it would also require a long seawall. Gray areas are the 100-year flood zones from before global fever.

The Thames Barrier is just downstream of Greenwich. This shows one of its smaller gates rotated into the upright position, blocking the storm surge from traveling farther upstream into London. The Thames Barrier is about 600 m wide; there is now a proposal to build a 16 km embankment, from Sheerness in Kent to Southend in Essex, containing similar gates to allow water to flow in and out of the Thames estuary.

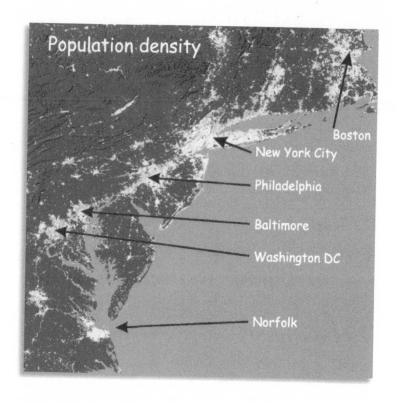

Population density (white dots) on the U.S. East Coast. The diagonal strip connecting Boston and Washington DC houses about one-fourth of the U.S. population. Economic disruptions in this corridor from hurricane (cyclone/-typhoon) storm surges would be widely felt elsewhere.

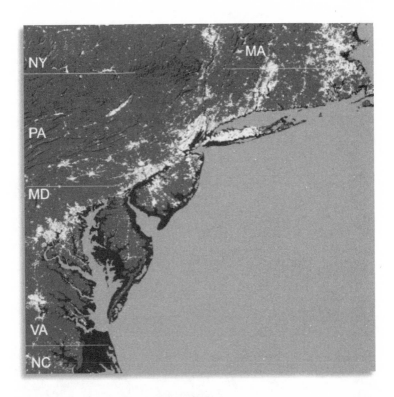

After a 6 m rise in sea level, many major population centers are partially flooded with saltwater (black areas). The Norfolk-Chesapeake-Newport area is underwater for about 50 km (30 miles) inland, continuing south into North Carolina and beyond.

That's Boston at top right; things don't look so bad on this scale of map.

But in fact eastern Massachusetts has a major problem when the sea rises 6 m/20 ft. The state capitol is still above it all on a new island, but MIT is underwater along with much of Boston. Harvard is waterfront on the new Harvard Peninsula (I used to live on its high point when I was at MIT and Harvard Medical School).

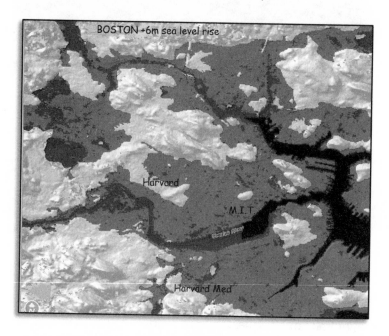

It isn't as bad on the west coast of North America because the shorelines rise more steeply. However, a mega version of San Francisco Bay will form, extending east into the Sacramento–San Joaquin delta. A large proportion of California's fresh water now flows through the delta; a salt disaster there would contaminate much of the drinking water for 32 million Californians and disrupt the world's seventh largest economy.

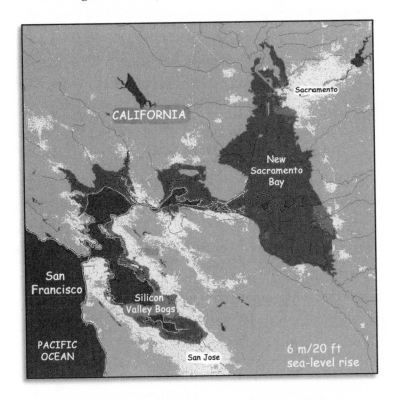

I have taken the privilege of naming some of the new marine features. New Sacramento Bay will be several times larger than the present San Francisco Bay. The Silicon Valley Bogs will enlarge San Francisco Bay to the south. Up north, some Napa Valley vineyards will become mangrove territory.

How soon can we expect the sea to rise by the entire 6 m? This century, the twenty-second, or just eventually?

Most of the twentieth century rise in sea level was from the thermal expansion of the oceans and the melting of many mountain glaciers. In the 2007 IPCC projection for the twenty-first century, the annual rise in sea level is calculated from the thermal expansion of the warming oceans, to which is added a small amount from the summer melt of the surface layer of the ice sheets. For later this century, the estimate came out to about 0.3 m, just half again as much as the twentieth-century sea-level rise.

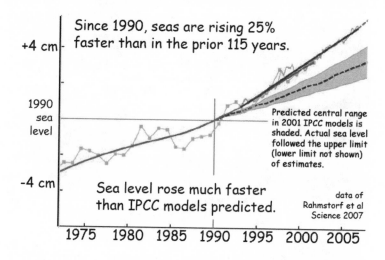

The Physical Sciences (WG 1) IPCC reports have a tendency to feature only what a giant spreadsheet can calc- ulate. They downplay aspects for which no firm numbers can be attached because mechanisms are not yet under- stood.

Many climate scientists would say that the IPCC reports are too conservative, that they underestimate the trouble ahead for the policymakers who rely on the reports. As an example, they point to the model in the 2001 IPCC report

which underestimated sea-level rise after 1990. And only
the models' high-end predictions for temperature can
match the actual data.

I suspect that the 2007 IPCC sea-level estimate of 0.3 m
in this century is completely inadequate as guidance for
policymakers. (A last-minute compromise was appended
to the IPCC report that allowed that another 0.4 m could
come from a faster melt of Greenland.) But 0.7 m doesn't
add up to the 3 m this century that many climate scientists
are worried about, just because of what's happened before.
This shows the eventual sea level rise for past temperatures:

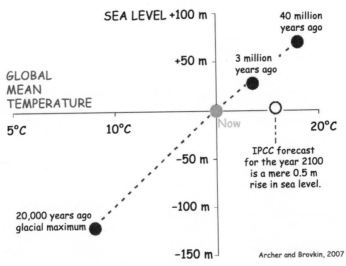

The diagonal line estimates the history of sea-level change as
global temperature slowly varies. It suggests that a 3°C fever will
eventually raise sea level by 50 m. That's only a hundred times
more than the IPCC forecast for this century.

To flood coastlines worldwide, ice sheets don't have to melt in place. They only have to melt enough to grease the skids. Think collapse, not melt. Most of the sea-level rise may instead come from ice being pushed off the land into the ocean. A little melting can quickly lead to a lot of sea-level rise.

It is said that a 6 m sea-level rise won't happen as quickly as Greenland's temperature rises by 3°C (which will only take a 1.5°C rise in global temperature, likely to occur this century). But we don't know how fast Greenland will collapse, having neither a comparable collapse in the climate records nor a tested dynamic model of collapse. It's clear, however, that the hills and valleys underlying an ice sheet can play a crucial role in slowing collapse, by pinning the bottom of the ice sheet.

The current data isn't good enough to say how fast it rose 125,000 years ago, but when the last ice age melted off, it rose as fast as 3 m per century. Of course, it was then warming up more slowly than now, as the Milankovitch factors slowly crept along to produce hotter summers at the latitude of the Canada's, Greenland's and Scandinavia's ice sheets.

That means that our projected global temperature rise of 2 to 6°C during the twenty-first century is at least ten times faster than back then—so we will likely see faster melt-offs in both Greenland and Antarctica as the Earth spikes a fever. We might even see collapse happen in new ways. I may hope that the sea-level rise is only 0.3 m this century as in the 2007 IPCC report, but it would be foolish to count on it.

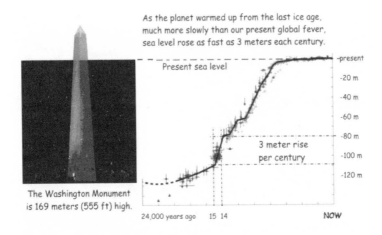

As the planet warmed up from the last ice age, much more slowly than our present global fever, sea level rose as fast as 3 meters each century.

Present sea level

3 meter rise per century

-present
-20 m
-40 m
-60 m
-80 m
-100 m
-120 m

The Washington Monument is 169 meters (555 ft) high.

24,000 years ago   15 14   NOW

Had the Washington Monument been built near sea level 22,000 years ago, the bottom 130 m would now be underwater. The worry is that the present rate, 0.3 m per century, will increase as Greenland collapses. The speed record, 3 m per century, is thought to have had a major contribution from the collapse of ice shelves in Antarctica.

However, our current warming is also proceeding ten times faster than back then and collapse mechanisms may be sensitive to this speed. So, while you should take seriously the estimates of sea-level rise itself (6 m/20 ft), all bets are off regarding how soon. No scientist is going to be able to say, "We're still safe for a century," even though we'll still see newspaper headlines like the Dutch one after the 2007 IPCC report was issued, saying "We're Safe!" —

based on, of course, that conservative sea-level rise estimate of 0.3 m.

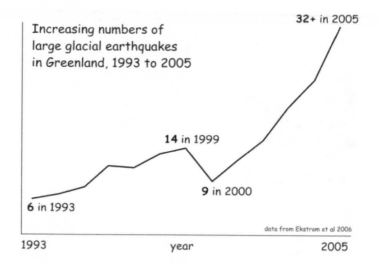

Increasing numbers of large glacial earthquakes in Greenland, 1993 to 2005

32+ in 2005

14 in 1999

9 in 2000

6 in 1993

data from Ekstrom et al 2006

1993   year   2005

Earthquakes are occurring under the major outlet glaciers from Greenland's central ice sheet. Quakes increased in the late 1990s—and that was followed by a rapid increase from 2002 onwards.

Perhaps they come from the ice briefly lurching forward and cracking, or perhaps breaking off protruding bedrock. Between the mid-1990s and 2005, their numbers doubled and then doubled again. In the same decade, ice flow measurements from radar showed that the yearly amount of ice exported as icebergs had doubled by one estimate, tripled by another.

If all this can happen with less than 0.8°C of warming since 1910, what will 3°C bring? Such dynamics only emphasize that the present computer models may under-estimate the trouble ahead. All that most people have heard about so far—if they've heard anything—has been based on the gradual extrapolation of present trends, not the more realistic scenarios. No one should count on the current melt being slow. When systems flip into new modes of operation, all bets are off.

I hope that it's obvious by now that serious sea-level rise is not controlled primarily by thermal expansion and by melting the ice surface at some predictable rate.

Rather, it is a matter of ice getting pushed off the land and instantaneously raising sea level. And how quickly ice gets pushed off the land depends on the ice mountain pancaking, spreading out sideways like that melting ice cream scoop on the sidewalk.

That in turn depends on what accelerates collapse, such as those water blisters that can form under the shoulders of the ice mountain, which create rotten ice and perhaps float the ice over obstacles.

This seems pretty obvious now, but in the IPCC reports, sea-level rise has been portrayed as slow, just one drip after another atop thermal expansion. It makes one wonder about how many other obvious things we are currently missing. Climate science is still a young science, akin to what neurophysiology was like in 1950 before positive feedback mechanisms were understood.

"May you live in interesting times" was an ancient Chinese curse, though I doubt that the collapse of Greenland or Antarctica was what they had in mind.

With structural famine gripping much of the subtropics [at a 2.5°C fever], hundreds of millions of people will have only one choice left other than death for themselves and their families: they will have to pack up their belongings and leave. The resulting population transfers could dwarf those that have historically taken place due to wars or crop failures. Never before has the human population had to leave an entire latitudinal belt across the whole width of the globe.

Conflicts will inevitably erupt as these numerous climate refugees spill into already densely populated areas. For example, millions could be forced to leave their lands in drought-struck Central American countries and trek north to Mexico and the United States. Tens of millions more will flee north from Africa towards Europe, where a warm welcome is unlikely to await them—new fascist parties may make sweeping electoral gains by promising to keep the starving African hordes out.

Undaunted, many of these new climate refugees will make the journey on foot, carrying what they can, with children and old people trailing behind. Many of them will die by the wayside. Uprooted, stateless, and without hope, these will be the first generation of a new type of people: climate nomads, constantly moving in search of food, their varied cultures forgotten, ancestral ties to ancient lands cut for ever.

—the writer Mark Lynas, 2007

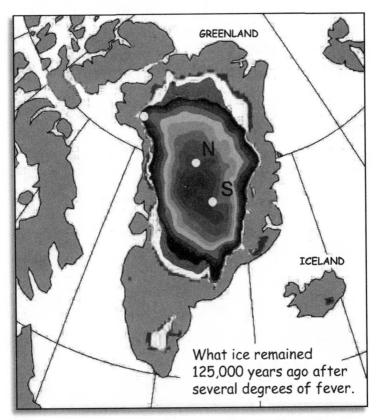

What ice remained
125,000 years ago after
several degrees of fever.

Nearly all of Greenland is presently covered by ice. The overlay shows the ice that remained after the last time that local temperature rose about 3°C . This occurred 125,000 years ago. **S** is the summit of Greenland where a pair of deep ice cores were taken in 1993; **N** is the site of another, slightly deeper, ice core.

Unfortunately, it only takes a 1.5°C rise in the global average temperature to produce a 3°C rise in Greenland—and 1.5°C is the smallest, most optimistic greenhouse warming estimated for the twenty-first century. The sea-level rise may take longer but it reaches more than 6 m / 20 ft. It may also happen more quickly than 2100 if enough ice slides into the ocean.

Greenhouse gas "natural cycles" and what's happened since 1850

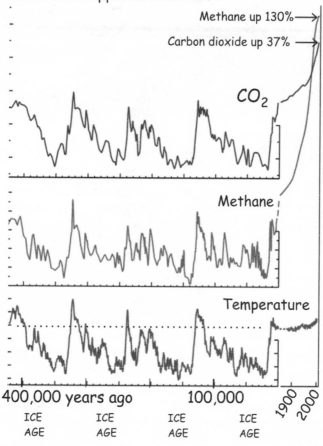

Methane up 130%→

Carbon dioxide up 37% →

CO₂

Methane

Temperature

400,000 years ago   100,000   1900 2000

ICE AGE   ICE AGE   ICE AGE   ICE AGE

When snow turns into ice, bubbles of air are trapped. Ice cores into the depths of Antarctica can sample 800,000 years of air, and thus determine how much CO2 and methane there was in the air. Because air mixes in only a few years, these are worldwide averages. (The temperature is that of the air in Antarctica when the snow fell; global mean temperature has different peaks.)

12

Methane Is the Double Threat

The [permafrost methane problem] does not have only a scientific character: it has passed to the plane of world politics. If mankind does not want to face serious social and economic losses from global warming, it is necessary to take urgent measures. Obviously we have less and less time to act.

—botanist Sergei Kirpotin, 2005

"It's all just some natural cycle." It might have been true, but it isn't. Contrary to what you might have heard, the present trends in heat-trapping gases are not just part of some natural cycle. They are evidence of our fossil-fuel fiasco.

There are now ice-core records from Antarctica going back 800,000 years for the major heat-trapping gases, allowing us to see their range of natural variation. For the last four ice ages, temperature has closely tracked methane (CH4 is "natural gas") and CO2 levels in the air.

To put the trends since 1850 in perspective, the scale is expanded at right. The two major heat-trapping gases have now gone far off scale. Temperature is beginning to follow (it would be much higher except for bright pollution from sulfur and ash masking a third of the expected rise in temperature).

Natural gas is mostly methane. It plays two roles in global fever. In its role as a fossil fuel that is burned, it produces the usual CO_2 as a by-product. But if the methane is not burned and leaks into the atmosphere, methane is 23 to 70 times more potent as a heat-trapping gas than CO_2.

There is naturally some methane in the air—indeed, methane vented from below the Temple of Apollo is probably the source of the trance and delirium of the Oracle at Delphi—but (see page 150) its concentration has risen about 130 percent over preindustrial levels, compared to 37 percent for CO_2. This is about a hundred times faster than anything observed in natural processes so far.

The alarm about methane came more recently than for CO_2, with Jim Hansen and Sherry Rowland warning of its increase in the early 1980s. Some climate scientists now worry that methane could become an even bigger actor on the global fever stage than CO_2 presently is.

{The] Mississippi was the only river in the world that had 'mud lumps.' [They] could rise suddenly enough to lift a ship as it passed. [They were] masses of tough clay, varying in size from mere protuberances looking like logs sticking out of the water to islands several acres in extent, [sometimes] three to ten feet [high]. Salt springs are found upon them, which emit inflammable gas.
—historian John M. Barry, 1997

Once upon a time, methane (CH4) mostly came from rotting vegetation in swamps ("swamp gas") and that buried under river deltas. Now we have anthropogenic methane as well: not just all that extra soil carried by the rivers from poor agricultural practices but methane from garbage dumps, rice paddies, cows that belch, pipelines that leak, and gas furnaces that emit some unburned methane every time they cycle on and off. (So do kitchen ranges and furnaces. That characteristic smell on startup is unburned natural gas.)

Over half of the gas given off by landfills is methane. At modern landfills, it may be captured and used for power generation. Thanks to the EU's Landfill Directive, much of the biodegradable waste stream that used to go to European landfills is being redirected to anaerobic digestion plants that deal with methane more efficiently. But in most of the world, landfills are releasing a lot of methane.

And then there is ancient methane. As with CO2, you can always distinguish ancient from modern methane by a version of radiocarbon dating. Methane occurs in assoc-

iation with coal and oil. It is what causes anoxia and explosions in coal mines. That's why canaries were watched to make sure they hadn't fallen off their perch. Mine methane is generally vented to the atmosphere. There are

still oil fields in the world which vent methane. More commonly, it is captured or burned off atop tall towers.

For the U.S. in 1996, about 1.4 percent of natural gas production was leaked, two-thirds of that from the pipes and compressors. No one knows how much the customers leak after the meter and the study excluded what is released during oil drilling and the like.

Most natural gas from the wellhead has substantial amounts of CO_2 mixed in. This has to be separated from the methane before use—and, of course, it is typically vented to the atmosphere. Norway and Algeria have pilot projects that pipe this CO_2 back underground.

Liquid natural gas (LNG) is now a booming business, being transported in a new generation of tankers to a new set of offshore ports that feed pipelines. LNG keeps itself cold by evaporation, adding methane to the atmosphere. Though the tankers may fuel themselves by capturing some of that methane, there are substantial losses of methane at every stage of the operation, especially in the land-based separation, compression, cooling, and (on the other end) rewarming processes.

In short, a quarter-century since the methane warnings, the industry is still building capacity as if unburned methane were not a potent greenhouse gas.

Then there's nature. The release of ancient methane from the ocean floor off California between Santa Barbara and the offshore islands is associated with periods of rapid rise in sea level. At a guess, the additional pressure of over-lying ocean might be squeezing out the hydrocarbons. If so, our greenhouse warming and sea-level rise could un-

leash fresh plumes of seabed methane, making the planet warmer and speeding the acidification of the oceans.

At the higher northern latitudes in Canada and Siberia, there is a lot of frozen ground beneath the surface. Some of it, especially in western Siberia, is from old lake bottoms littered with decaying vegetation. When the ground thaws, it releases a lot of methane and CO_2. This escapes into the air and adds to the insulating blanket around the Earth.

Thaw lakes expand like potholes in the Siberian permafrost. (Photograph by methane researcher Katey Walter in 2003.)

This, of course, produces a vicious cycle. Global fever (twice as high in the Arctic) thaws more permafrost, which releases more methane and adds to the greenhouse effect.

Though methane is the least offensive of the fossil fuels, in terms of electricity generated per unit CO_2 released, it is the most dangerous of the three fossil fuels when unburned. Coal adds twice as much CO_2 to the atmosphere per megawatt, but at least it doesn't evaporate.

In a rational world, we'd burn methane only near its wells, sinking its CO_2, and exporting the electricity via efficient DC transmission lines that can reach long distances. Even better, retired natural gas pipelines could be re-used as conduits for superconducting power lines.

For the long-term, there's another concern about methane. There's a lot of it held in an ice-like form called a

The ice that burns, methane hydrate.

methane hydrate, often buried under the ocean floor or slope. It sometimes comes unstuck and pops up to the surface. Sailors will fish out a piece and amuse themselves by setting it afire.

It is thought that huge amounts of methane were released about 55 million years ago, perhaps by earthquakes causing an avalanche on an underwater slope.

But there are signs of progress. While the methane concentration in the air was growing at the rate of 100 ppb per decade, something slowed the rate of addition so that, in 2000 to 2006, it matched the rate of decay (losing methane for making CH3 methyl groups). Before a problem gets better, it has to stop getting worse—and perhaps the methane contribution to global fever is about to head down.

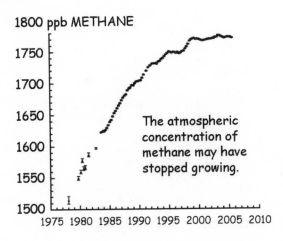

The prior high for methane, about 320,000 years ago, was half of the 1978 value.

The only good thing you can say about our methane problem is that methane doesn't hang around in the air for as long as CO_2. We will lose about half of this year's methane additions in the next 6 years, which makes it much faster than the CO_2 decay rate (we lose half within

200 years but that's the easy half; most of the remaining half will still be around a thousand years from now unless we start removing it from the atmosphere). Unlike methane, we're going to have to remove CO_2 rather than wait it out.

No one really knows why the methane release has slowed, though there have been many steps taken that are candidates. If it really is a turnaround and methane is on the way back down, it might be the second greenhouse-gas success story after ozone (back at page 27).

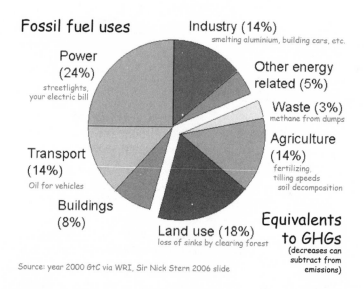

Source: year 2000 GtC via WRI, Sir Nick Stern 2006 slide

Since CO_2 is not the only problem, what number should we really pay attention to? What best expresses the damage done? The climate modelers use a technical term called "forcing" but I find that this pie chart does a good job. It shows that two-thirds of the insulation problem comes from fossil fuel uses. What's interesting is the remaining third.

There are a series of equivalents to adding CO_2, such as taking away CO_2 sinks. The biggest one is the loss of sinks by land clearing. This piece of the pie (mostly thanks to Indonesia, with Brazil far behind in second place) is even bigger than oil's use for transportation. Agriculture contributes another 14 percent by tilling the soil (speeds decomposition), fertilizing (nitrous oxide), burping cattle, and all that waste. About 3 percent of the pie comes from garbage dumps and the like.

The global climate models of the past have not taken much account of the positive feedbacks that exaggerate temperature rise from fossil fuel CO_2 uses. A 2006 analysis reports that

> Increased photosynthesis at higher CO_2 levels and temperatures implies a negative feedback, but positive feedbacks seem likely to override this effect. For instance, higher temperatures may lead to increased release of CO_2, methane and N_2O from terrestrial ecosystems...

For example, the soil decomposes faster when warmer, one reason why tropical soils are so poor, compared to those that stay frozen in winter. Decomposition releases CO_2 and methane into the air.

Such feedback increases the warming already "in the pipeline" up to 1.5°C. And it amplifies the effect of doubled CO_2 from an additional 1.5°C to as much as 4.5°C.

That's why runaway warming can occur at some point. That's when we lose existing sinks for carbon dioxide via drought, fire, ocean acidification, narrowing the leaf pores, and baking the soil. Losing sinks works just like adding CO2, pushing up the fever to kill off even more sinks. The temperature spike just keeps building up, totally disrupting much of life on Earth. While no one is predicting a result like Venus, the Earth's outgassing would be a profound extinction event.

The CO2 runaway scenario is perhaps the biggest danger that we are currently flirting with.

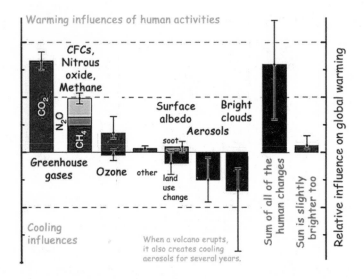

The largest anthropogenic (human-caused) actor is CO2. The net forcing (the difference between inflows and outflows) was thirteen times greater for human-linked effects than that from the sun's brightening. So while the sun is a player, the score now stands at:

```
┌─────────────────────────────┐
│          SPONSORING         │
│        GLOBAL FEVER         │
├──────────────┬──────────────┤
│    HUMANS    │     SUN      │
├──────────────┼──────────────┤
│              │              │
│      13      │      1       │
│              │              │
├──────────────┴──────────────┤
│       FOURTH QUARTER        │
└─────────────────────────────┘
```

As you come through the door of a supermarket, a unit above your head blasts you with hot air in the winter and cold air in summer (sometimes, when the manager has not been paying attention, it is the other way around). You must stand blinking for a moment as your eyes adjust to the [bright] lights.

Then you walk past banks of fridges and freezers which have no doors. This would be impossible to believe, if it were not by now one of the most ordinary facts of life. But, though you walk through valleys of ice, you remain warm.

All day long, the freezers and the heaters must fight each other. They must do so in a building which is huge, generally uninsulated and often widely glazed: that is capable, in other words, of trapping neither heat nor cold.

—George Monbiot, 2006

Two stable states.
Enough of a push and the ball
gets trapped in the upper valley.

The other stable state: what 113 mph (182 km/h) winds did to many trucks along the interstate highway between Idaho and Utah in 1999. (Photo by Marta Storwick in the *Standard-Examiner* of Ogden, Utah.)

13

Sudden Shifts in Climate

It's the extremes, not the averages, that cause the
most damage to society and to many ecosystems.
— climate scientist Claudia Tebaldi, 2006

Most of us are familiar with the concept of a tipping point.
When everything is precariously balanced, you can tip
either way. Tipping back may allow a safe landing. Tipping the other way may cause something more dramatic,
perhaps a crash, slip, or flip.

In a flip, you leap from one stable state to another, and
very quickly (think about a light switch, popping from *off*
to *on* with a snap). What happens to a big truck being
rocked by a strong wind is that it can either return toward
its typical stable state (upright) or crash into another stable
state (flat on its side). Crash examples so far include Snowball Earth, the collapse of the food chain by El Niño, and
deep drought burning off the Amazon rain forest.

You've probably heard of the abrupt climate change
caused by the shutdown of the northern offshoot of the
Gulf Stream. It's happened many times in the past, about

every few thousand years during the last ice age, so we had to assume we were vulnerable to getting hit by another. But the scientists who make working models of the ocean circulation have now been able to show that, without the extensive winter sea ice, another abrupt shutdown would be unlikely.

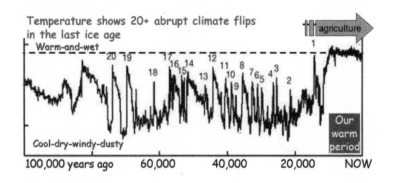

Indeed, the next ice age also appears to have been indefinitely postponed—all thanks to how much our fossil-fuel fiasco has warmed the earth. In just a decade after I wrote "The great climate flip-flop" for the *Atlantic Monthly*, there has been a lot of scientific progress.

The sinking of surface waters in the Greenland Sea declined during the last half century and is forecast to continue declining throughout the twenty-first century—but, say the modelers, it is not likely to drop in a mere five years.

The real question, to my mind, is whether the past abruptness was due to the winds rearranging themselves because of the altered map of sea-surface temperatures that results from the sinking slowdown. For the future, that might matter. A decade of ocean current decline might be the setup for a year of abrupt wind rearrangement.

Rearranging the customary winds is an obvious possibility for the climate popping into a new mode of operation. Global warming is not uniform. Because of all the evaporation, the equatorial regions might warm only 1°C at the same time that, thanks to all of those lighter-to-darker feedbacks, the Arctic warms much more. Winds are driven by temperature differences, so it won't surprise climate scientists if the winds change.

This need not happen gradually over the years. Indeed, in an El Niño, it only takes a few months.

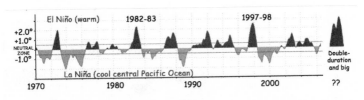

The trade winds in the tropics blow out of the northeast in the Northern Hemisphere, and out of the southeast in the Southern. As such, they converge near the equator to flow west. This pushes a lot of warm surface water across the Pacific, making sea level higher in the western Pacific than near South America. Some years, the trade winds become weak and fickle. That's what starts an El Niño episode.

Normally it is very hot and humid in the western Pacific just northeast of New Guinea, thanks to that "warm pool" from the trade winds. Evaporation shifts into high gear, with air and moisture carried upward in great thunderheads.

What goes up must come down. In the equatorial Pacific, that air usually comes down dry on the Americas side of the Pacific. This forms an east-west circulation pattern known as the Walker Cell (seen at the top of page 167).

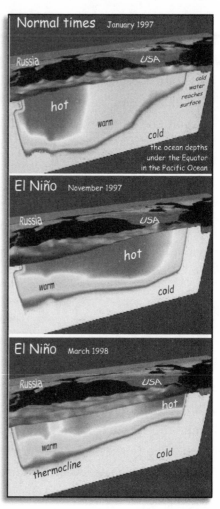

Normal times   January 1997

Russia   USA

hot

warm

cold water reaches surface

cold

the ocean depths under the Equator in the Pacific Ocean

El Niño   November 1997

Russia   USA

hot

warm

cold

El Niño   March 1998

Russia   USA

hot

warm

cold

thermocline

In an El Niño year, lacking those strong trade winds, the western warm pool stays in the central Pacific Ocean. Lots of hot air rises from a new place. This causes the Walker Cell to split, with a new downdraft developing to the west, creating dry winds—and, in the 1997–98 El Niño, causing forest fires and famine in Southeast Asia and New Guinea. The eastern loop shifted east from the Galapagos Islands and dried the Amazon Basin.

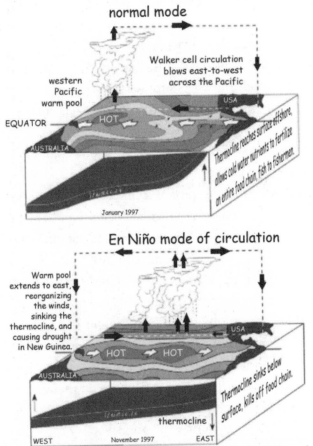

normal mode

Walker cell circulation blows east-to-west across the Pacific

western Pacific warm pool

EQUATOR

HOT

AUSTRALIA

USA

Thermocline reaches surface offshore, allows cold water nutrients to fertilize an entire food chain, fish to fishermen.

January 1997

En Niño mode of circulation

Warm pool extends to east, reorganizing the winds, sinking the thermocline, and causing drought in New Guinea.

HOT   HOT

USA

AUSTRALIA

Thermocline sinks below surface, kills off food chain.

thermocline

WEST   November 1997   EAST

With global fever, we could stay stuck in the El Niño mode.

Looking north at a cross-section of the Pacific Ocean along the equator. There is normally a pool of relatively hot water north of New Guinea and south of the Philippines. Near South America, the ocean surface stays relatively cool. During an El Niño, the trade winds weaken and so the hot pool spreads east. This rearranges the winds worldwide, with dry air descending over the rainforests of southeast Asia and the Amazon basin. Drought is followed by forest fires.

This new mode of atmospheric circulation took several months to develop and it lasted for a year. There is much concern that global fever will create a "permanent" El Niño situation, of the type seen in the Earth's climate records from a million years ago.

Even in the usual El Niños, there are rippling effects. The only good one is to suppress hurricanes en route to Florida and the Gulf of Mexico (that's what stopped the 2006 hurricane season). El Niño also rearranges storm tracks across the U.S., often causing flooding in the South and drought in the north.

But when the equatorial winds are rearranged, it reduces the upwelling of nutrients from the ocean depths offshore of the west coasts of North and South America. The ocean depths are laden with nutrients that sank. Carrying them up to the near-surface fertilizes photo-synthesis in the phytoplankton. More plankton means more little fish, and eventually more big fish and fisher-men. Too few remaining nutrients can cut back on primary production so far that it kills off the food chain in many places. That's why the offshore fisheries on the west coasts of North and South America are so sensitive to the re-arrangement of the winds by El Niño.

The other change is in sea level itself. Sea level is only level during an El Niño!

Normally, the Pacific Ocean is a little tilted because the strong trade winds near the equator push a lot of water westward, raising sea level near the Philippines (see top panel of figure on p.167). But during an El Niño the trades

east from the Galapagos Islands and dried the Amazon Basin.

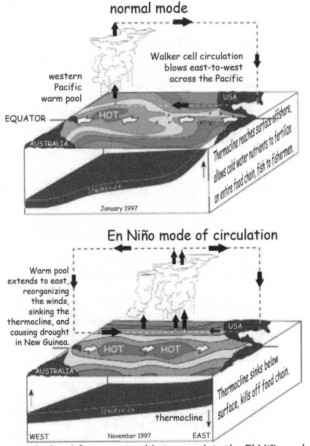

With global fever, we could stay stuck in the El Niño mode.

Looking north at a cross-section of the Pacific Ocean along the equator. There is normally a pool of relatively hot water north of New Guinea and south of the Philippines. Near South America, the ocean surface stays relatively cool. During an El Niño, the trade winds weaken and so the hot pool spreads east. This rearranges the winds worldwide, with dry air descending over the rainforests of southeast Asia and the Amazon basin. Drought is followed by forest fires.

This new mode of atmospheric circulation took several months to develop and it lasted for a year. There is much concern that global fever will create a "permanent" El Niño situation, of the type seen in the Earth's climate records from a million years ago.

Even in the usual El Niños, there are rippling effects. The only good one is to suppress hurricanes en route to Florida and the Gulf of Mexico (that's what stopped the 2006 hurricane season). El Niño also rearranges storm tracks across the U.S., often causing flooding in the South and drought in the north.

But when the equatorial winds are rearranged, it reduces the upwelling of nutrients from the ocean depths offshore of the west coasts of North and South America. The ocean depths are laden with nutrients that sank. Carrying them up to the near-surface fertilizes photosynthesis in the phytoplankton. More plankton means more little fish, and eventually more big fish and fishermen. Too few remaining nutrients can cut back on primary production so far that it kills off the food chain in many places. That's why the offshore fisheries on the west coasts of North and South America are so sensitive to the rearrangement of the winds by El Niño.

The other change is in sea level itself. Sea level is only level during an El Niño!

Normally, the Pacific Ocean is a little tilted because the strong trade winds near the equator push a lot of water westward, raising sea level near the Philippines (see top panel of figure on p.167). But during an El Niño the trades weaken, allowing the tilted sea surface of the Pacific to relax back toward being level.

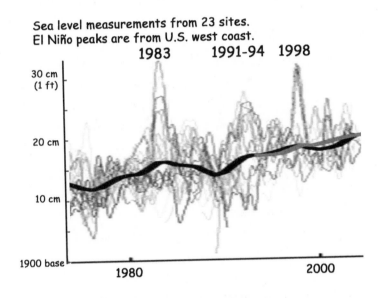

Sea level measurements from 23 sites. El Niño peaks are from U.S. west coast.

Smokestack and tailpipe emissions of CO_2 may be what have caused most of our problem, but it's the balance of CO_2 sources and sinks that we care about. That's because all CO_2 sources and sinks can function to restore CO_2 to normality.

Consider a lake. Its level is set by what sources (rain, upstream river) add and what sinks (leaks, evaporation, downstream river) take away in the course of a day. You can draw down the reservoir by reducing inputs or by increasing outputs. The balance point depends on how sinks vary with, for example, lake depth and therefore water pressure on any deep outlets.

For climate, the number of sinks may increase slightly from the fertilization effects of CO_2 (though higher concentrations act to limit uptake). The higher temperatures, however, may promote soil decomposition and fires, both adding sources and taking away sinks.

Before 1750, when most heating and cooking involved burning wood and dung, sources and sinks balanced out when CO_2 was at about 280 molecules out of every million (ppm). And with burning trees, most of the carbon was recycled the same year back into young plants. But then we began taking fossil carbon out of long-term storage and dumping it into the air. Since the sinks were not increased by an equal amount, the CO_2 concentration is now rising, causing global fever and ocean acidification.

We think of forests as reliable carbon sinks. But they are only net sinks, and can turn into net sources if rotting wood emits more CO_2 than the remaining leaves can take up. Soil is also full of decaying organisms. If baked, it releases more and more CO_2. The tropical rain forests can reverse from being sinks to being sources. It happens during an El Niño.

And, of course, even rain forests can burn down, returning all of their stored carbon to the atmosphere.

That pushes up the fever and further accelerates the decay in top soils, which add more CO_2. This accelerates the temperature spike. This is why we worry about a climate runaway. The heat of, say, 2040 may be sufficient to guarantee that 2050 will be even hotter—whatever we do during the decade. That's a climate runaway.

It's also called trashing the planet and creating a mass extinction event. Fortunately, there are more reliable ways to sink CO2.

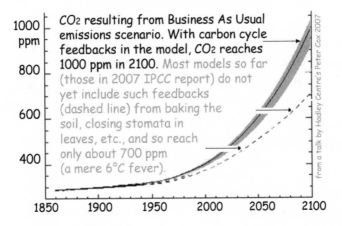

CO2 resulting from Business As Usual emissions scenario. With carbon cycle feedbacks in the model, CO2 reaches 1000 ppm in 2100. Most models so far (those in 2007 IPCC report) do not yet include such feedbacks (dashed line) from baking the soil, closing stomata in leaves, etc., and so reach only about 700 ppm (a mere 6°C fever).

from a talk by Hadley Centre's Peter Cox 2007

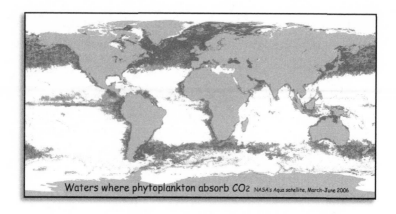

Waters where phytoplankton absorb CO_2 NASA's Aqua satellite, March-June 2006

The oceans have tiny algae called phytoplankton that remove CO_2 from the air and release oxygen. The darkened areas show the March-to-June amounts of chlorophyll in surface waters.

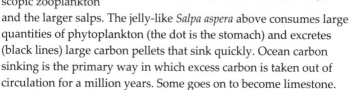

The phytoplankton feed the microscopic zooplankton and the larger salps. The jelly-like *Salpa aspera* above consumes large quantities of phytoplankton (the dot is the stomach) and excretes (black lines) large carbon pellets that sink quickly. Ocean carbon sinking is the primary way in which excess carbon is taken out of circulation for a million years. Some goes on to become limestone.

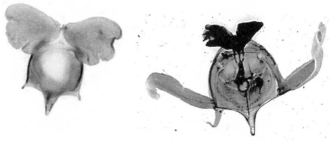

The microsnails above, both pteropods ("winged feet"), have a shell. When they die, that sinks some more carbon. Several decades ago, Victoria Fabry discovered that the pteropods do poorly in excess CO_2.

14

A Sea of CO$_2$

It's even worse than "You don't know the half of it," at least when it comes to climate change. *An Inconvenient Truth* may have prepared you for the bad news about what's happening on land. But the other half of the bad news is what happens at sea.

I was shocked when I went directly from the bird-rich Galapagos Islands down to Easter Island, and discovered there were few shore birds and not much there that they could have fed on. I asked the Chilean archaeologist who was showing me around the island. He said the algae had been killed off by the warmer ocean, and with them the zooplankton that eat algae, next the fish, and then the birds farther up the food chain.

The big El Niño episodes had done this and water temperature has remained warm enough to inhibit recovery. The local fishermen have been reduced to seeking deep water fish, not very common, difficult to hook, and requiring expensive gear.

Jim Hansen and colleagues reported in 2006 that warming over the past century has been greater in the western Pacific than in the eastern. They suggest that the increased west-east temperature difference may have increased the likelihood of strong El Niños, such as those of 1983 and 1998.

Staghorn coral, bleaching

A coral reef beginning to "bleach" (the plant part is ejected from the animal part) following environmental stress, such as over-heating for too long.

Episodes of excessively warm water are also what have been killing off the coral reefs in many places. Vast areas of coral shed their gaudy coloration, turn bone white, and die if the heat continues for weeks. The 1983 El Niño was the first large-scale bleaching event in at least 300 years. The one in 1998 destroyed 16 percent of the world's coral reefs.

The hot summer of 2005 raised water temperatures all along the typical hurricane tracks into the Caribbean and the Gulf of Mexico. Besides paving the way for such devastating hurricanes as Katrina and Rita, the hot water

directly affected many coral reefs. Marine biologists in
Puerto Rico reported that 42 coral species on some reefs
had bleached (areas of mere bleaching are not shown on
the map). Coral colonies more than 800 years old died in a
matter of weeks. It was worse farther east in the Virgin
Islands. Coral that survives initially may die from
subsequent disease; divers later saw die-offs as far down
as 90 ft (27 m).

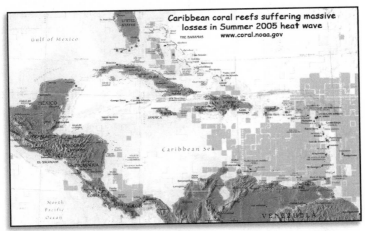

The shaded areas above show the areas where it remained hot
enough for long enough to bleach and then kill corals. Many other
areas (not shown) bleached.

The Great Barrier Reef off Australia is already in serious
trouble. The twenty-first century may well see the world's
last coral reef die, say some of the experts.

Atmospheric CO_2 has two parallel effects, global warming
and ocean acidification. From each, there is a fan-out of
impacts. In the atmosphere, elevated CO_2 produces warm-

ing—and warming in turn may kill people (35,000 Europeans died in the heat wave of 2003), diminish cereal crops, expand the subtropical deserts, set up long-lasting droughts elsewhere, and cause the largest species extinction event since the demise of the dinosaurs.

In the ocean, it may be even worse, though the science is much less well established than on land. It wasn't until about 2003 that the magnitude of the ocean problems began to alarm scientists. My colleague Ed Miles said, "We are making changes in ocean ecosystems—changes not seen for millions of years—and we don't know what will happen. We just don't know."

We do know part of what is already happening. About 85 percent of all the extra heat captured by the CO_2 blanket has been taken up by the oceans. It is reported that the southern oceans may have absorbed about as much CO_2 as they can.

More than half of the oxygen we breathe comes from photosynthesis by the near-surface phytoplankton and microalgae. Some of the microscopic animals in the sea (zooplankton, such as the diatoms pictured) that eat them grow tiny exoskeletons, taking up the carbon from the CO_2. When they die (and this happens quickly because of "bloom and bust"), these "shells" sink to the ocean depths. Some become limestone. Other zooplankton excrete carbon pellets which sink. This "biological pump" is presently the major pathway for taking excess carbon out of circulation for millions of years. As such, it is far more reliable than growing forests that can burn down, quickly putting the captured carbon back into the air.

Compiled from 'The Diatoms' by Round, Crawford and Mann (1990)

Diatoms and coccolithophores also sink carbon when they die.

Waves carry air bubbles down into the surface waters. Before a bubble rises to the surface and pops, its CO_2 starts influencing the carbon chemistry in the ocean around it. A sea of CO_2 not only reduces pH ("ocean acidification") but diminishes this "carbon pump" in more direct ways. It cuts the supply of the carbonate ions that combine with calcium ions to form the compound calcium carbonate, used for building shells and coral—and then limestone.

Normally the sea water has more than enough carbonate but when the elevated CO_2 drives the carbonate equilibrium far enough out of balance, this starts pulling calcium carbonate out of shells and coral—tearing down rather than building up. The zooplankton then start looking malformed and dysfunctional. Both the coral and the calcite glue that holds a reef together get into similar trouble, dissolving like a cube of sugar. By dissolving coral reefs, the CO_2 adds to more global warming by removing an important carbon sink.

On land, some additional CO_2 in the air can serve as a fertilizer for some crops. But in the ocean, high CO_2 acts as an herbicide (indeed, it is used to kill all of the plankton in a tanker's ballast water before it is dumped, to avoid introducing new species in distant places).

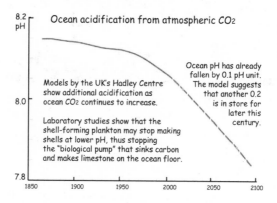

The important conclusions already reached by the researchers in the field: we must cut carbon emissions and pump down the CO_2 concentrations in the atmosphere, not merely cool down the earth by some geo-engineering project. Our fossil CO_2 is hitting at the bottom of the ocean food chain. And it is diminishing our most important carbon sink, just the sort of thing that could produce a runaway warming condition.

Acidification of seawater can cause high mortality rates in a variety of fish species when they are in their larval stage and part of the zooplankton. Bad as these effects of CO_2 are, there is an even more serious one. The total number of phytoplankton ("primary production") in the world's oceans have been on a decline.

> I think there's a whole category of organisms that have been around for hundreds of millions of years which are at risk of extinction—namely, things that build calcium-carbonate shells or skeletons. To a first approximation, if we cut our emissions in half it will take us twice as long to create the damage. But we'll get to more or less the same place. We really need an order-of-magnitude reduction in order to avoid it.
> —climate modeler Ken Caldiera, 2006

A "bloom" of phytoplankton occurs when sufficient nutrients are brought together, say near a river mouth or sewage outfall. But winds may suffice. A steady wind that pushes aside the surface waters can bring cool underlying waters to the surface, carrying along sunken nutrients from the depths.

In many places, plankton's ability to pump down CO_2 is limited by the loss of building materials. When the shell-forming zooplankton die and sink, they take with them the calcium, phosphates, and other nutrients needed by the next generation.

One theory for the widespread decline of plankton is a lack of fertilization by the iron that is carried off the red deserts by high winds. In the cool-dry-windy-dusty climate of the ice ages, the atmospheric CO_2 drops down to its baseline, 180 ppm. This is widely attributed to iron fertilization.

Intense
phytoplankton
bloom in 2006

BRITISH
COLUMBIA

WASHINGTON

COLUMBIA RIVER plume

OREGON

Wind-driven upwelling of nutrients caused this large phytoplankton bloom offshore of Canada and the U.S. in 2006.

The westerlies in the southern oceans create an entire band of upwelling circling the globe. (See page 172.)

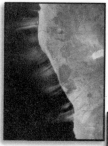

Large sand dunes along the coast of Angola and Namibia regularly fertilize the offshore blooms with iron dust carried by the winds. At Walvis Bay (beneath large plume at the bottom), there are sea lions everywhere, carpeting most of the beaches, attesting to the

rich food supply for the top of the food chain.

Dust plumes from Iceland have been seen fertilizing large blooms of phytoplankton.

GREENLAND

ICELAND

Chlorophyll imaging shows
2007 phytoplankton bloom

dust plumes
from Iceland

So, to naturally remove some fossil CO_2, why not create a bloom on demand, simply by spreading around some iron dust? There have been a number of "iron-enrichment" experiments, yielding valuable data on single applications of iron to enhance phytoplankton production.

The zooplankton bloom usually doesn't occur until a week or so later. Not much research has yet addressed that follow-on bloom—indeed, it is only some of the zoo-plankton species which are relevant for long-term sinking of carbon. Fortunately, some French scientists have begun to study a "natural experiment" where iron enrichment continues for many months over an area the size of Switzerland.

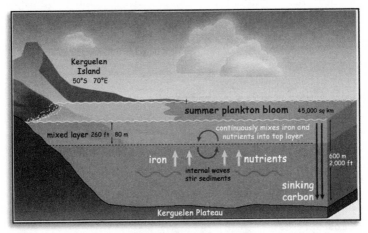

Kerguelen underwater plateau, to the southeast of the volcanic island in the Indian Ocean, has internal waves that stir up the iron in the sediments. This keeps the annual bloom going for a few months, sinking ten to a hundred times more carbon, per unit of iron, than was estimated during the previous single-application studies. The Kerguelen bloom quits when there is no longer enough iron.

Iron fertilization around the edges of such a Switzerland-sized bloom would be an obvious strategy because such waters are near the threshold for a bloom already. Some coccolithophore species have even larger blooms. In 1991 south of Iceland, *Emiliania huxleyi* had a bloom three times the size of Iceland.

I suspect that terrestrial carbon sinks are going to prove unreliable because of drought, heat waves, and fire. We already have extensively cleared areas that were once forested. The largest such area is the southern Amazon Basin in Brazil, though Canada is similarly black in the diagram below.

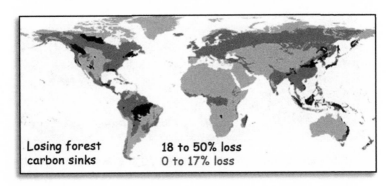

Losing forest carbon sinks — 18 to 50% loss — 0 to 17% loss

Given that 70 percent of the earth's surface is water, we're likely to try out some schemes to solve our need for more carbon sinks. I intend the following scheme only as a brief hint at a plankton management technology, what we might see later in this century.

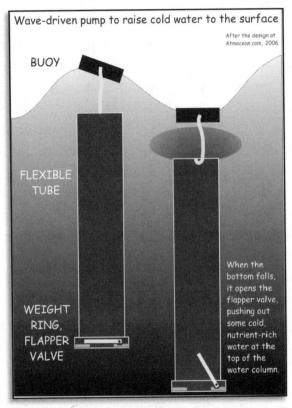

Wave-driven pump to raise cold water to the surface

After the design at Atmocean.com, 2006

BUOY

FLEXIBLE TUBE

WEIGHT RING, FLAPPER VALVE

When the bottom falls, it opens the flapper valve, pushing out some cold, nutrient-rich water at the top of the water column.

Technology can perhaps mimic the natural upwelling from the nutrient-rich depths in less well endowed regions of ocean surface. Inventors have been busy patenting schemes, despite the lack of money for demonstration projects. One patent is simple enough to explain in one page. I wish I'd thought of it myself. The diagram shows a simple scheme for using the ups and downs of waves to power the uplift of cold, nutrient-rich deep waters.

Coming down off the top of a wave, the tube falls with the flapper valve open. After some hours, deep water arrives at the top of the column. Now when the weighted column drops after a wave crest, deep water is forced out at the top of the column. The higher the wave, the greater the pump stroke.

Imagine thousands of such columns around the edges of a natural bloom such as Kerguelen's, extending the bloom's area by providing extra nutrients from the deeper waters. And the waves are always high at Kerguelen.

That's my speculative example for what might be in store, once we learn a lot more about safe plankton management. Because the flapper valve can always be latched closed, the managers of an anchored array of hundreds of such tubes could turn nutrient upwelling on and off. Near such managed arrays would become a good place to fish.

Its inventors point out that an array of up-tubes could cool surface water by a degree or so—which, if done in the typical hurricane/typhoon/cyclone tracks, could reduce wind strength before landfall.

Note that waves can also pump surface waters down. Just place the flapper valve at the top, facing down. Down-tubes provide another method of sinking the dissolved CO_2 and carbon-laden nutrients in the surface waters, as well as some of the heat.

Serious scientific warnings about rising CO_2 started in 1956, so we have a half century of history to illuminate why the societal response has been so sluggish. Science serves as our headlights and, if the applications of technology outdrive the headlight reach, it may prove impossible to stop in time. Most of our climate problem comes from very simple technologies (axe, plow, drill) but without the high-tech science satellites, we would be blind to major changes. Without the working models of climate, we would have little idea of how sensitive the climate mechanisms are to small changes.

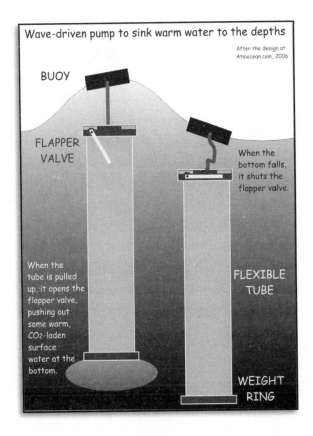

Wave-driven pump to sink warm water to the depths

After the design at Atmocean.com, 2006

BUOY

FLAPPER VALVE

When the bottom falls, it shuts the flapper valve.

When the tube is pulled up, it opens the flapper valve, pushing out some warm, CO2-laden surface water at the bottom.

FLEXIBLE TUBE

WEIGHT RING

But even with good data on how global warming and its effects have been occurring for the last fifty years, even with good coupled circulation models for atmosphere and ocean to show us the consequences, society has mostly ignored the increasingly emphatic warnings. There are partial exceptions such as Europe and the state of California, but there are also many global actors trying to modernize rapidly. Some actors (U.S., Australia, Canada) are addicted to high-energy (and high-garbage) lifestyles and have been, despite their advanced technological abilities, unwilling to take even baby steps toward conservation or carbon-free ("C-free") energy.

Avoiding tipping points does not come for free. Civilizations have good reflexes only if they build them in. We have not.

Fatalism takes it for granted that we are not masters of our fate. But, while that is obviously true in some sense, it does not follow that god-will-provide is the correct attitude to adopt (what most people associate with fatalism). Our capabilities are quite different than when those predictable Greek tragedies were written 2,500 years ago.

Most civilizations in the past have proven fragile. We're the first one to understand what's going on. We're likely capable of repairing the rot we have caused in the foundations. But it takes a great deal more. People pay much more attention to political and religious leaders—and to actors—than they do to scientists, so much depends on whether they help to lead an effective response.

While denial and deception have played roles in our slow response, there is nothing here that seems peculiar to intelligent life on Earth. Explosions in population and consumption seem likely to be found in any society where intelligence is not mature enough to head off such problems. One suspects that, for every galactic society that solves its problems in time, there will be hundreds that snuff out their own candle.

For decades, the U.S. has been the moral, economic, and military leader of the free world. What will happen when we end up in Planetary Purgatory, facing 20 or more feet of sea-level rise, and the rest of the world blames our inaction and obstructionism, blames the wealthiest nation on earth for refusing to embrace even cost-effective solutions that could spare the planet from millennia of misery? The indispensable nation will become a global pariah.
— oceanographer Joseph Romm, 2007

But these people [of hard-hit countries] may not be content to remain passive victims, for they will surely know that the world they inherit is not one that they have created. The resentment felt by Muslims towards Westerners will be tame by comparison. As social collapse accelerates, new political philosophies may emerge, philosophies which seek to lay blame where it truly belongs — on the rich countries which lit the fire that has now begun to consume the world.
— the writer Mark Lynas, 2007

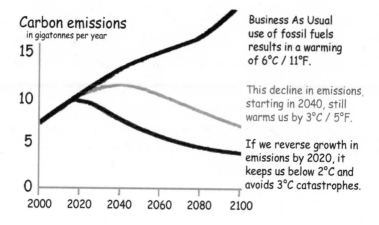

Carbon emissions
in gigatonnes per year

15

10

5

0

2000   2020   2040   2060   2080   2100

Business As Usual use of fossil fuels results in a warming of 6°C / 11°F.

This decline in emissions, starting in 2040, still warms us by 3°C / 5°F.

If we reverse growth in emissions by 2020, it keeps us below 2°C and avoids 3°C catastrophes.

The emission of greenhouse gases ... is causing global warming at a rate that began as significant, has become alarming and is simply unsustainable in the long term. And by long term I do not mean centuries ahead. I mean within the lifetime of my children certainly; and possibly within my own. And by unsustainable, I do not mean a phenomenon causing problems of adjustment. I mean a challenge so far-reaching in its impact and irreversible in its destructive power, that it alters radically human existence.... There is no doubt that the time to act is now.

—British Prime Minister Tony Blair, 2005

15

The Extended Forecast

For I dipped into the future,
far as human eye could see,
saw the vision of the world,
and all the wonders there would be.
—Alfred Tennyson, 1842

Everyone loves to complain about the weather. And, of course, about weather forecasts that aren't perfect. This long tradition of authority-bashing for weather forecasts can carry over to the public's opinion of whether climate forecasts should be taken seriously. If this were about, say, the prospects for an influenza pandemic, there would be far less second-guessing of the experts.

Climate is the longer-term overview of weather. It consists of the averages and the extremes of hot and cold, wet and dry, snowpack and snowmelt, winds and storm tracks, ocean currents and upwellings—and the long and short of it, their patterns in space and time. The paleo-climate records from long cores of ocean bottoms, ice sheets, and bogs have been very useful for telling us what can happen and how fast.

About 38 years of Greenland ice layers from 16,250 years ago, a time when the dust storms in Asia left their mark in Greenland (darker bands).

What will happen if the earth's fever climbs even more? The simplest way of judging is to consult past history for analogous situations. For example, were the earth to warm up several degrees more, is there data from a past warming to suggest what could happen?

Yes. The last time that the earth warmed up more than present was only 125,000 years ago, during the warm interglacial before our most recent ice age. It warmed Greenland by 3°C, what a general global warming of 1.6°C would do this century. As I mentioned earlier, this melted a lot of ice in Greenland and Antarctica, enough so that the oceans rose more than 6 m / 20 ft above today's sea level (see the figure on page 120).

So there's your first-order estimate: if it works like it did last time, just the untreated fever expected before midcentury will be enough to set in motion a large rise in sea level.

Besides the history of temperature change, you can use working models to project the Earth's current fever into the future—models run on a computer, rather than in a lab mockup. Scientists like to run experiments where you change something and see things play out differently.

To forecast changes in annual rainfall does not require forecasting the rain each day for a year (remember that for the next time you hear "Why, they can't even get the weather forecast right for the weekend—why should we believe them about the year 2050?"). These models are not the weather forecast models which try to predict the next few days. Changes in annual precipitation only requires knowing the basic physics of temperature and pressure, then making a working model of the winds that should occur, then of the dew point where water vapor turns into fog or clouds.

Climate scientists have been making global-circulation models for both the ocean and the atmosphere. Dozens of labs around the world are competing to discover mistakes and have improved the state of the art over the last three decades. These global-climate models have been a great success.

A model will generate trade winds and westerlies, with dry zones between them, without being told to do so. If something like a volcanic eruption comes along, the model shows the expected cooling for several years. Such a model is not detailed enough to separately represent every cubic meter of the air or ocean, and so they cannot be used to model hurricanes in any detail. Most do not slice up time finely enough to do day after day weather forecasts.

The initial modeling focus is on the known past. If the model's CO_2 concentration is forced to match the Keeling curve, does the resulting temperature and rainfall follow what was actually observed? The models have gotten pretty good at it.

It's only at this point when the model is taken seriously as showing us what future climate might be. The best of these models are used for the IPCC reports. Typically the models are run many times, using the world weather on different days to start up the run.

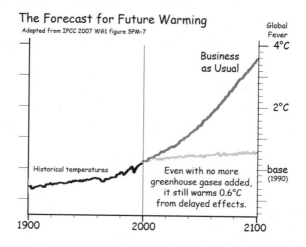

The forecast for future air temperature, °C averaged globally. Double it for the interior of continents or higher latitudes; almost double again for °F. For scale, warming up from an ice age is a 5°C shift over nine centuries—but we may produce a similar shift within one century. For some processes, the rate of warming may matter more than the temperature itself.

These climate models predict unusually strong hurricanes like Katrina, fiercer heat waves, harsher droughts, heavier rains, and rising sea levels as global warming intensifies. And, of course, the models give you the degrees of fever to be expected.

The climate model forecast for twenty-first century warming depends on what happens with future fossil-fuel use. The hottest scenario typically graphed, Business as Usual, should not be considered an extreme case. It's for world economic expansion much like today's pace except that population growth is assumed to peak at midcentury.

Our interference with climate is already at dangerous levels in terms of floods, droughts, wildfires, species loss, and melting ice, just from our 0.8°C fever at the end of the twentieth century. Carrying on as usual carries enormous risks, condemning today's students to a world of constant insecurity and frequent catastrophes.

And we're already "committed" to a higher fever, just from the delayed effects of twentieth-century carbon emissions. That lowest curve assumes that greenhouse gases remain at their values for the year 1990 (in other words, that fossil-fuel use simply stopped or was cancelled by new carbon sinks) and it still shows a 0.6°C rise in the fever.

This much fever is not inevitable, of course, as we could increase sinks enough to counter the "commitment" if we tried hard enough to remove CO_2 from the air and sink it.

Climate change has a very high procrastination penalty that just grows and grows with each passing year of inaction—rather like what happens if you don't pay off your credit card. But for climate, there is no such thing as a fresh start from bankruptcy.

"Climate sensitivity" measures how strongly the Earth's climate system responds to a given perturbation—say, a volcanic eruption—and is often expressed as the equilibrium rise in global temperature resulting from doubling the pre-industrial CO_2 concentration (275x2=550).

The baseline for temperature (no "fever") is simply the 1990 global mean temperature. There's also nothing special about doubling; it's just a convenient benchmark.

The conventional value for sensitivity is about 3°C, but it might be twice as great—warm a degree, get one free. So permit me a brief digression about how we know that. The graph at right shows the fever which results for five different values of the peak atmospheric CO_2 concentration. The center bar, whose indicated midpoint is the oft-quoted 3°C fever, is produced by model runs that all end up settling down at 550 ppm of CO_2.

But that's only the most common result. About 95 percent were cooler than 6°C (11°F) average global fever; 95 percent were higher than a 2.4°C fever. The reason for a range of sensitivities is that the model is re-run many times with slightly different starting values for, in effect, the weather conditions on the first day of the simulation. Ninety percent of the results lie between 1.6° and 4.7° with the average being a 3°C fever.

Nature, however, doesn't do this averaging to get 3.0°C and the actual sensitivity might be one of the outliers that yields a 6°C fever—merely because Mother Nature got started on the wrong foot.

There are some ways to study the ice cores that ought to yield an independent estimate of climate sensitivity. By the time of the next big IPCC report, climate scientists might be able to narrow down the uncertainty in climate sensitivity.

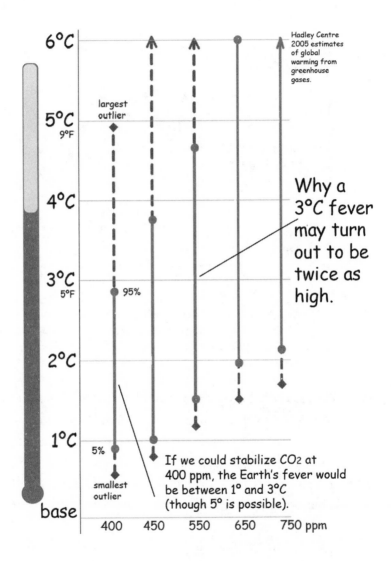

Hadley Centre 2005 estimates of global warming from greenhouse gases.

Why a 3°C fever may turn out to be twice as high.

If we could stabilize CO₂ at 400 ppm, the Earth's fever would be between 1° and 3°C (though 5° is possible).

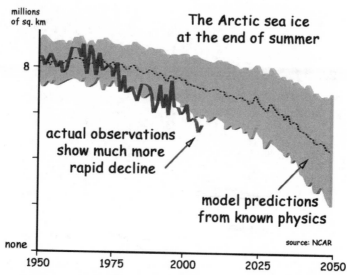

The actual decline in Arctic sea ice was about three times faster
than the IPCC models would have predicted. One possibility is
that soot was underestimated. Another possibility is that the 2007
IPCC sensitivity (3°C rise for a CO_2 doubling) may underestimate
the fever track that we're on.

Since there is sure to be a lot of quibbling about "alarmist"
climate forecasts, let me outline why scientists have come
to view the matter so seriously. It's no longer a matter of
one theoretical line of argument, as it was in 1898. It's no
longer a matter of two correlated lines of data, CO_2 and
global temperature, the way it was in the half century that
followed.

Dozens of things have gone wrong with climate over
the last fifty years that fit the theoretical framework for
fossil-fuel climate change. Even if there were an alternative
explanation for several of those—say, land use changes
proving more important for the fifty-year trends in

wildfires and floods than global warming—it wouldn't change things in the other areas. Nor would it change the forecasts for the future.

So let me explain where the climate change conclusions come from. Major scientific and medical conclusions start with the individual investigators, usually working on one piece of the puzzle. They submit their data and conclusions to peer review. An editor selects two to five other experienced investigators in the same field to read the manuscript, not only to "grade the paper" but to write a report on it. This often results in conclusions being toned down; sometimes it forces more data to be collected to fill the holes identified by the reviewers, whereupon the rewritten manuscript is resubmitted for more peer review.

Once published in a peer-reviewed journal, any important conclusion will not stand alone for very long because other researchers will be attracted to the subject and soon there will be a whole cloud of peer-reviewed papers on the subject. Usually, the original conclusions will be qualified or extended.

Important topics will soon require a scientific assessment, where an expert panel of scientists takes a careful look at all the evidence and writes an evaluation. This meta assessment is itself peer-reviewed, then published. For example, the entire literature on weight-loss dieting was recently evaluated in this manner. (Diets fail on four out of five tries, and the health hazards of a medically unsupervised diet may well swamp whatever benefit remains. This will, of course, alarm the profitable $35-billion diet-foods industry.)

Climate science does scientific assessments (themselves peer-reviewed) every few years. However, the major international scientific assessments of climate change and its impacts have an additional twist or two, as the science historian Spencer Weart explains:

> The Intergovernmental Panel on Climate Change . . . was created by conservatives [in the early 1990s] to forestall "alarmist" declarations from self-appointed committees of scientists. Governments committed the IPCC to repeated rounds of study and debate, forbidding any announcement except by unanimous consensus. It seemed a sure formula for paralysis.
> However, the power of democratic methods, combined with rational argument, overcame all obstacles. The IPCC has evolved into a robust transnational institution that provides authoritative conclusions of grave significance.

So for the physical science portion of the 2007 IPCC report, thousands of scientific findings were reviewed, 600 climate scientists were involved in writing the report, then it went out to peer reviewers, then to additional national reviewers like me—so thousands of reviewers (making a total of 30,000 comments) were involved before it ever landed in the laps of hundreds of government representatives gathered in Paris—who proceeded to tone down some well-established scientific conclusions. Fortunately, someone will always compare the draft report with the finished one to show the changes, put up the comparison on the web, and the countries that pushed the changes then become publicly identified with trying to rewrite the scientists.

To call something "alarmist" that survives such a winnowing process is to risk one's own credibility. It is more likely that the IPCC reports understate the case, that the alarm has been somewhat muffled by the cumbersome process.

As mentioned earlier, the sea-level rise for the twenty-first century in the IPCC report is mostly thermal expansion because they were uncertain how much to add for accelerated iceberg production. The models used for the 2001 IPCC report predicted CO_2 rise correctly but were on the low side for both temperature and sea-level rise.

Remember also that the IPCC estimates leave out anything that can't be assigned a reliable number—and so they leave out quite a lot, including history. A big part of IPCC's problem is its strict adherence to the use of physical models. By IPCC standards, "if it's not in a model, it's speculation," says Stefan Rahmstorf, one of the leading modelers. By ignoring factors that can't yet be modeled, he says, IPCC came up with deceptively reassuring numbers.

The other omission is the whole class of sudden episodes of damage, as when the Amazon burns in a prolonged El Niño.

The culture of science (in contrast to medicine) downplays the possibility of things going badly wrong. Mumbled British understatement is a style not confined to UK scientists. Jim Hansen calls it "scientific reticence." As the physicist Mark Bowen said, "Scientists have an annoying habit of backing off when they're asked to make a plain statement, and climatologists tend to be worse than most." Graduate students will witness enough examples of someone being shot down for "overstating the case" that they will become cautious, adopting their professors' style of understatement. Except in a few areas such as engineering and medicine, risk assessment usually isn't part of the scientific culture. It's one of the reasons why we've been getting "climate lite" reports for so long.

As we decide what to do about climate change, it is well to remember the standards in other areas for making "expensive" decisions. A U.S. jury in a criminal case uses the standard of "beyond a reasonable doubt" because we, as a society, have decided that convicting an innocent person by mistake is unacceptable. In a civil case, where only money and reputation is at risk, the standard is relaxed to "a preponderance of the evidence." As I learned serving as a jury foreman in both civil and criminal cases, it is much easier for a jury to agree in a civil case.

People also take sensible precautions when the risk is high. Ask a roomful of people if they have fire insurance. Almost all will raise a hand. Ask how many have had a fire in the last ten years, and almost none will respond. Yet people pay for insurance because, should a fire happen, they could lose everything—and still have to pay off the mortgage.

The 2006 Stern Report estimates the annual cost of climate insurance at below 5 percent of GDP (though the economy would take an annual 20 percent hit if action is delayed). Furthermore, much of the carbon-free ("C-free") remake is soon going to be needed anyway because both oil and natural gas production will decline. This means that much of the makeover's expense is a cost we will soon bear anyway, even if a miracle prevented further global warming.

We need to innovate in a hurry and, considering our reputation for technological innovation, it's odd that the U.S. has fallen behind.

Many U.S. geothermal installations are imported from Israel. Though the Hot Rock Energy concept was invented in 1972 at Los Alamos, the action is now in Europe and Australia.

The U.S. hasn't started a nuclear plant since 1978. The French and Japanese are now the most experienced. The French electricial utility will help build, and is a major investor in, the U.S. nuclear plants in the pipeline.

The Germans have become the innovators in "green" buildings.

The Japanese and the Germans have more experience with solar power panels, though innovative thin-film photovoltaic from the U.S. is starting to make a difference.

Think of wind turbine innovations and Spain, Denmark, and Germany will come to mind.

To my surprise, there are now Italian

windmills back home in Kansas! The largest Italian utility company is building a big wind farm near Hayes, Kansas to sell clean electricity to the coal-dependent American Midwest.

Countries that innovate early get the new jobs, developing an economic edge over the C-free laggards that end up having to later import the technology.

> I spent a day discussing climate change at No. 10 Downing Street [in 1989, Prime Minister Margaret Thatcher gave her Cabinet a seminar on global warming], and sitting next to me was Mr. James Lovelock, the author of the Gaia theory. When we went down to the street afterward there were lots of journalists waiting, and they all thrust microphones into our faces. They asked Lovelock, "What do you expect to happen after today?" He just said, "Surprises."
>
> —British diplomat Sir Crispin Tickell, 2002

Speaking of surprises, the 2002 National Research Council report on abrupt climate change had the subtitle: *Inevitable Surprises*. One reason we have a standing army is in case we are surprised. Now we must provide something similar for climate surprises. We need the resilience to bounce back when something unexpected hits.

We also must have a good safety margin. We routinely have safety margins in support strength or fire resistance. No architect would design a stadium without calculating the weight of the fans packing those bleachers—and then, for safety, doubling the number. Actually, it's the noncritical components that have a safety factor of two. For components whose failure could result in substantial financial loss, serious injury, or death, a safety factor of four or higher is common. Yes, that's more expensive, one reason why building collapses are common in countries where building inspectors can be bribed to ignore skimping on the materials.

For climate protection, we need a safety margin in schedule. Our response needs to make a lot of progress up front, just as insurance against something as unexpected as, say, a fire in a rain forest.

We've only got one habitable planet and we dare not shave our margins.

It is now entirely plausible that about a rise of 1.5$^{\circ C}$ globally will mean the end of coral reefs and polar bears. That about 2$^{\circ C}$ will mean catastrophic melting of Greenland and Antarctic ice, with commitment to multi-meter rises in sea level. That about 2.5$^{\circ C}$ will sharply reduce global crop yields.

Thus stopping at twice the pre-industrial CO_2 (550 ppm corresponding to about 3$^{\circ C}$, once thought a reasonable target by many) may not be good enough.

Many analysts and groups now conclude that prudence requires aiming not to exceed 2$^{\circ C}$.

—climate scientist John Holdren, 2006

Begin with 100 units of primary energy—coal, say. In generating electricity in a typical power plant, 66 units of energy go right up into the sky as waste heat. For automobiles the waste is even worse. It's as if, when you cooked a meal, you were to take two-thirds of the food and immediately dump it into the garbage.

Then, when we use the electricity, there is further waste, as when inside a lightbulb, heat is made along with light. To continue with the meal analogy, it's as if, after you served the food (the one-third that was left), each diner threw away 90 percent of that. Wouldn't that be appalling? You'd have to say there was something wrong with any food plan that operated this way. In other words, about 3 percent of the energy in the fuel gets turned into light.

—physicist and playwright Phillip F. Schewe, 2007

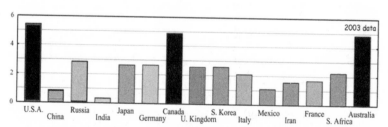

An American, Australian, or Canadian uses twice as much fossil fuel as a European.

Lifestyles that consume twice as much energy (and produce twice as much garbage) per person are characteristic of the two developed countries to reject the Kyoto treaty, the U.S. and Australia (and, under a conservative government, Canada also made such noises). In 2007 data, Saudi Arabia joined the big-black users club.

16

Doing Things Differently

I recently attended a post-performance discussion of Thornton Wilder's play, *By the Skin of Our Teeth*. It won the Pulitzer Prize for Drama in 1943 and features an ice age in Act 1, a deluge in Act 2, and ends with the aftermath of a great war. Almost everyone in my discussion group saw the play from the context of our present climate crisis. They seemed in despair over what could be done about it.

I'd say that they understood the essentials of our climate crunch but could not yet imagine how such problems could be solved. In the next four chapters, I will try to remedy that.

Individually, we have trouble imagining how to go on an energy diet that cuts our carbon consumption in half. Many people feel fenced in by a series of time and money constraints, with little wiggle room. But what is problematic for individual initiative can often be accomplished by high-level energy policy and collective action.

The purpose of this short chapter is simply to show, via comparisons between states and countries, that much has already has been done for cleaning up electricity generation. We can do a great deal by simply copying the strategies of the successful.

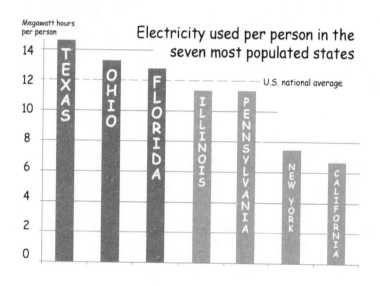

For example, Texans use more than twice as much electrical power per person as do Californians. They will say that this is because they need to heat and cool more. (Really? More than New Yorkers who, despite their overheated buildings, are almost as frugal as the people of California?)

For some insight, take a look at how this situation developed over a half century. Consumption per person tripled in the United States on average, but it only doubled in California, thanks to how California changed course in the 1970s. By going their own way, the people of California have kept electricity consumption per capita flat for the last thirty years.

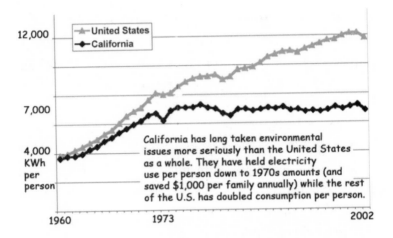

California has long taken environmental issues more seriously than the United States as a whole. They have held electricity use per person down to 1970s amounts (and saved $1,000 per family annually) while the rest of the U.S. has doubled consumption per person.

As this shows, it can be done without a drastic diet. California is nonetheless thriving; were it a country, it would have the sixth largest economy in the world. "California's experience shows that serious conservation is a lot less disruptive, imposes much less of a burden, than the skeptics would have it," the economist Paul Krugman wrote. "And the fact that a state government, with far more limited powers than those at Washington's disposal, has been able to achieve so much is a good omen for our ability to do a lot to limit climate change, if and when we find the political will."

Why can't Texas be more like California? The California Solution shows that good government policy can create a framework within which market forces can operate.

My second comparison, which I call the Nine Percent Solution, comes from looking around the world at how electricity is generated. In France, nuclear and hydro supply all but 9 percent of France's electricity. It appears that minimizing fossil fuels can also be done while maintaining an excellent standard of living.

Because much of the U.S. fossil fuel supply has to be imported, many dollars have to be exported. This dollar drain often goes to governments that are not our friends; our oil payments have financed major terrorist movements directed at us and our friends.

Even without the global fever problems, this U.S. energy policy has been a disaster, much like what our worst enemies might have wished on us if they wanted to weaken us or slowly take us over. A country which owes so much money to Arab and Asian countries is not likely to retain an independent foreign policy for much longer.

Hundreds of millions of dollars have been spent—successfully, to judge from the lack of Congressional investigations over the years—to rig the system and convince Americans to trust their future to those who sell fossil fuels.

Trusting your energy policy to the fossil fuel lobby is like trusting your health care system to the tobacco lobby.

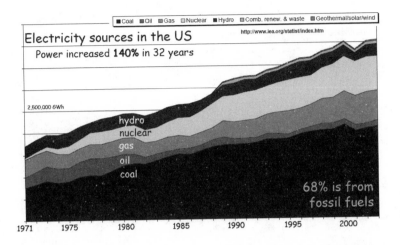

The growth in electricity use in France has been twice as steep as in the U.S. Yet France now gets only 9 percent of its electricity from fossil fuels. (Graphs adapted from those at *www.iea.org-/Textbase/stats/graphsearch.asp.*)

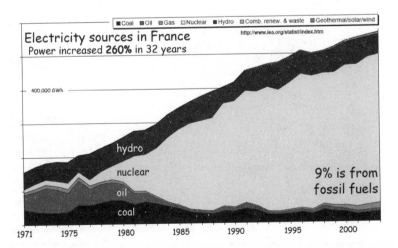

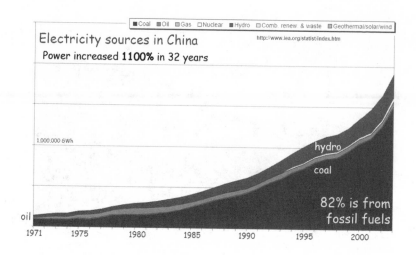

Electricity sources in China

Power increased **1100%** in 32 years

http://www.iea.org/statist/index.htm

1,000,000 GWh

hydro

coal

82% is from
fossil fuels

oil

China's use of coal has soared; nuclear power is that light sliver
starting in 1993 atop the gray oil sliver. The UK has replaced some
of its oil and coal use with Russian natural gas starting in 1992,
thereby reducing its CO_2 contributions.

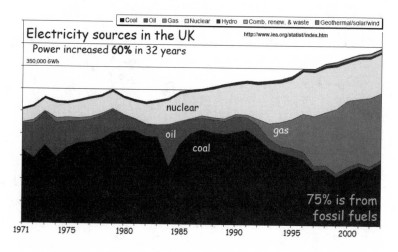

Electricity sources in the UK

Power increased **60%** in 32 years

http://www.iea.org/statist/index.htm

350,000 GWh

nuclear

oil   gas

coal

75% is from
fossil fuels

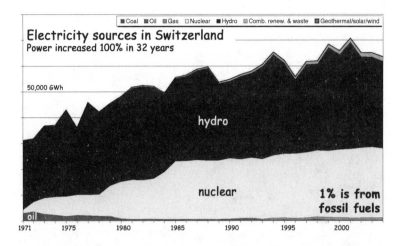

Switzerland added nuclear to supplement hydro and now has 99 percent clean energy. Their total is actually decreasing, presumably from efficiency gains. India, despite enormous hydro resources and many nuclear engineers, is choosing to burn coal to meet electricity needs.

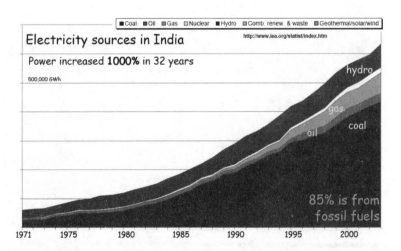

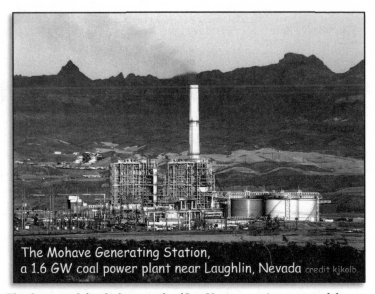

The Mohave Generating Station,
a 1.6 GW coal power plant near Laughlin, Nevada credit kjkolb.

This huge coal-fired plant south of Las Vegas, a major source of the haze over the Grand Canyon, sits atop some of the best geothermal resources in the U.S., in the midst of good sites for wind farms, with year-round sunshine for solar energy.

One coal-fired power plant requires three mile-long coal trains every day.

I'd put my money on the sun and solar energy. What a source of power! I hope we don't have to wait until oil and coal run out before we tackle that.

—inventor Thomas Edison, 1931

Governments... spend a small slice of tax revenue on keeping standing armies, not because they think their countries are in imminent danger of invasion but because, if it happened, the consequences would be catastrophic. Individuals do so too. They spend a little of their incomes on household insurance not because they think their homes are likely to be torched next week but because, if it happened, the results would be disastrous. Similarly, a growing body of scientific evidence suggests that the risk of a climatic catastrophe is high enough for the world to spend a small proportion of its income trying to prevent one from happening...

The real difficulty is political. Climate change is one of the hardest policy problems the world has ever faced. Because it is global, it is in every country's interests to get every other country to bear the burden of tackling it. Because it is long term, it is in every generation's interests to shirk the responsibility and shift it onto the next one. And that way, nothing will be done...

Developing countries argue, quite reasonably, that, since the rich world created the problem, it must take the lead in solving it. So, if America continues to refuse to do anything to control its emissions, developing countries won't do anything about theirs. If America takes action, they just might.

— from the *Economist*, 9 September 2006

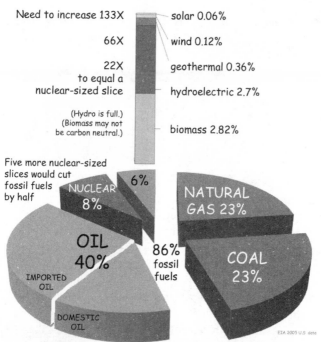

Need to increase 133X — solar 0.06%

66X — wind 0.12%

22X
to equal a
nuclear-sized slice — geothermal 0.36%

hydroelectric 2.7%

(Hydro is full.)
(Biomass may not
be carbon neutral.) — biomass 2.82%

Five more nuclear-sized
slices would cut
fossil fuels
by half

NUCLEAR 8%

6%

NATURAL GAS 23%

OIL 40%

86% fossil fuels

COAL 23%

IMPORTED OIL

DOMESTIC OIL

EIA 2005 U.S data

U.S. energy (all uses): **Where it comes from**

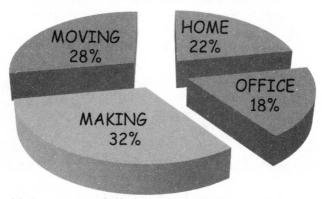

MOVING 28%

HOME 22%

OFFICE 18%

MAKING 32%

U.S. energy (all sources): **Where it goes**

This conventional breakout ignores most of the waste heat.

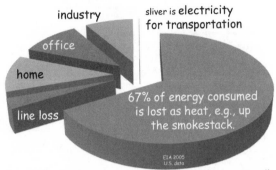

industry   sliver is electricity for transportation

office

home

line loss

67% of energy consumed is lost as heat, e.g., up the smokestack.

EIA 2005 U.S. data

Electricity's use of energy (70% wasted)

At least for electrical generation, about 70 percent of the fuel's energy is wasted as heat. A small fraction of electrical generation's waste heat is currently captured ("Co-generation") for local heating needs, thus reducing some electrical demand. Much oil is wasted because of idling time in traffic jams.

The southern end of a low-loss 1,500 km DC electrical transmission line just outside Boston. All of the DC-to-AC conversion is done in that small building. The transformer farm is for the AC en route to Boston.

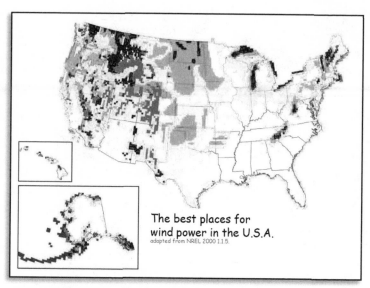

The best places for
wind power in the U.S.A.
adapted from NREL 2000 1.1.5.

Among the best places for wind power is out in the middle of the
Great Lakes. There are many windy sites offshore as well (below).

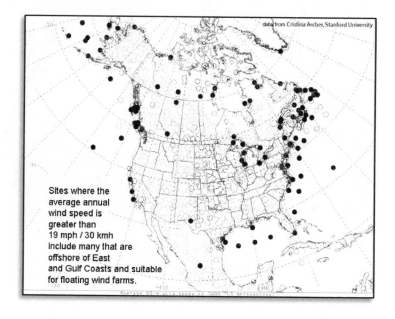

data from Cristina Archer, Stanford University

Sites where the
average annual
wind speed is
greater than
19 mph / 30 kmh
include many that are
offshore of East
and Gulf Coasts and suitable
for floating wind farms.

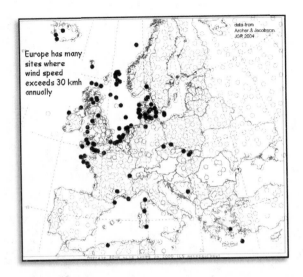

Some C-free electricity sources have serious drawbacks when scaled up too fast. That's because the power grids themselves are already unstable, as frequent power outages remind us. That they don't happen more often is only because most sources and loads are predictable. When unpredictable wind and solar power is added to the balancing act, network outages become more likely.

The current projection for growth in wind has it replacing 5 percent of fossil fuels by 2020 and that may be as fast as we should go unless we use energy storage such as a flywheel or an underground reservoir for compressed air. Some cities in Iowa are planning a 100 megawatt wind farm with 200 megawatts stored underground as compressed air.

My sunrise photograph is of the two refineries in Anacortes, Washington, with Mount Baker in the background near the Canadian border. The refineries consume 255,000 barrels of Alaskan crude oil per day, delivered by supertankers, and produce gasoline, propane, and diesel fuels.

17

Cleaning Up Our Act

Time is short and the prospect of even partial success remains uncertain. Yet we can avoid catastrophe by mobilizing our ingenuity and community spirit.

Addressing global warming will require less sacrifice than defeating fascism and communism, but more foresight—and that is exactly what we have been acquiring. If humanity's track record with long-term problems shows mostly indifference and failure, that need not set a precedent for our future.

—historian of science Spencer Weart, 2007

I am often asked whether I think we're up to acting on climate in the United States, given that the glaciers in Greenland are moving faster than our politicians.

The tone of the question is usually pessimistic, but my answer is not. I can imagine scenarios that might prove effective, given some political geniuses rising to the challenge.

I reply that even the best governments find it difficult to handle problems that require a lot of expert opinion to be translated into appropriate action with major economic consequences. Still, there are success stories.

The U.S. Congress has, in the past, created commissions to which they have delegated some of their powers. A non-technical example is the commission for closing military bases. This is a delegation of powers which saves a senator or representative from being directly linked with the loss of a big payroll back home. But I'm thinking more of the long-standing commissions requiring expertise, professionalism and experience, where debate, discussion, and deliberation are tempered by the oversight that a functioning democracy expects and demands.

For example, the Securities and Exchange Commission makes rules that promote transparency. While this is more about managing the playing field than it is the stuff of science and technology, little would get done if Congress hadn't handed off some of its powers. The Federal Communication Commission is similar but more technical (just imagine Congress trying to deal with spread-spectrum frequency allocations).

The Federal Reserve Board is the best-known and, as Stanford climate scientist Stephen H. Schneider has noted, the most relevant model for action on climate. Not only does "the Fed" save the Congress from having to vote every three months on unpopular issues like raising interest rates, but it brings economic expertise to matters that go beyond what wisdom Congress might collectively possess on the subject.

> In theory, you could design a tax and rebate system [which, though it] would be no harder to implement than a rationing system, it would, because of the complex system of fees and rebates, be more difficult to explain. Complex ideas seldom do well in politics, as most people do not have the time or patience required to understand them. We are likely to react against one part of the package before we have grasped the whole idea.
> —commentator George Monbiot, 2006

In the U.S., it is clearly time for Congress to institute European-style fossil carbon taxes while reducing other taxes to compensate people for paying higher fuel bills. (Cap-and-trade will raise prices without a reduction elsewhere.)

At present, there is little that the average person can do to reduce their exposure to U.S. taxes. I like Al Gore's notion of eliminating payroll taxes (social security/-Medicare and unemployment; they're the biggest part of withholding for most people) when a carbon fee kicks in to put a price on pollution. This is really clever social engineering, not mere tax relief. Presently there are few ways to game the system and pay less taxes than your neighbor. But with the C-fee (a better phrase than carbon tax) in place, then a more efficient car, better insulation, and car-pooling work like tax credits, not deductions. People love to game the system and the prospects of reducing their total taxes by a third will bring out all sorts of creativity that will reduce carbon emissions and grow more C-sinks.

Second, it's time to delegate substantial rule-making power to an expert commission for climate policy. For example, the Federal Carbon Board might adjust carbon tax rates and emission caps to ensure that national CO_2 and smog goals are met, just as the Federal Reserve Board now adjusts the mortgage and credit card interest rates to ensure that inflation and unemployment targets are met.

Third, we'll need a Carbon "Makeover" Commission to mandate more efficient cars, trucks, planes, buildings, appliances, and manufacturing processes. Some of the C-fee money needs to support the development of the longer-term technologies, things that no company can currently undertake while remaining competitive. The commissioners will also need to quickly build demonstration projects such as geothermal plants (more in Chapter Nineteen). They will need to make sure that oil and coal companies do not buy up the alternative fuel companies and then sit on their innovative patents until the clock runs out. This makeover opportunity offers the largest, cheapest, and fastest leverage on carbon emissions—which is why Congress cannot be left to deal with it, piece by piece.

Fossil fuel use has been growing annually and, if it continues to grow at the same rate (rather than even faster), it will double worldwide before midcentury. The Carbon Initiative researchers at Princeton show what it will take to replace this growth by dividing the efficiency and replacement problem into seven areas. Each "Princeton wedge" is assumed to improve fast enough to replace 1 billion tons per year by midcentury.

The wedging strategy for stopping the annual increase in fossil carbon burn rate
modified from Sokolow et al (2004)

14 billion tons per year of carbon emitted

currently projected path of fossil carbon per year

7 wedge stabilization triangle

Wedge #1 flat path constant yearly emissions until mid-century

7

historical growth of yearly carbon emissions

doubled wedges drive carbon use back down to zero

mid-20th C   NOW   mid-21st C

For example, wedge #1 might come from doubling gasoline mileage, #2 from reducing distances driven by half. Replacing all incandescent light bulbs in the world with compact fluorescents might provide one-fourth of a wedge. One or more wedges might come from increased use of geothermal and nuclear power plants, and so forth.

This framework has no actual reduction of carbon emissions until after midcentury—and stopping the annual increase in carbon emissions (which is all it addresses) is a long way from stopping the increase in CO_2, or the main thing, actually removing the damaging CO_2 from the air. Stopping emissions growth represents the most minimal of do-something responses. But the exercise is useful because it creates a way to "compare apples and oranges."

In my opinion, we cannot afford to proceed so slowly. But even a doubled wedge doesn't actually reduce the CO_2 in the atmosphere, only the rate of its increase. Nor does the wedging strategy provide a safety margin for, say, regional failures to participate—or a big double-duration El Niño whose fires bump up the CO_2 and therefore the global temperature.

> Avoiding temperature increases greater than 2–2.5°C would require very rapid success in reducing emissions of methane and black soot worldwide, and global carbon dioxide emissions must level off by 2015 or 2020 at not much above their current amount, before beginning a decline to no more than a third of that level by 2100.
>
> —the UN's Scientific Expert Group on Climate Change and Sustainable Development, 2007

Carbon reductions take a long time to show up because the CO_2 from every supertanker and coal train stays in the atmosphere for centuries. The other contributors to global fever can, when reduced, show an effect more promptly, suggesting major efforts ought to be directed at them in light of the deadline of 2020.

The world's anthropogenic methane comes from energy systems and livestock (each about 30 percent), 25 percent from agriculture, and another 15 percent from landfills and waste treatment. That's good news in the sense that technology can reduce all of them. Methane's relatively short atmospheric lifetime, 12 years, means that improvements translate quickly into reduced concentrations and reduced global fever.

The soot that warms the atmosphere comes from the old two-stroke engines and the common diesel engines, as well as from traditional burning of wood and dung, agricultural burning, and forest fires. The emissions from engines and biomass fuels can be sharply reduced by technical means. The short atmospheric lifetime of soot (days to weeks) means that reducing it has immediate effects on global fever and restoring the rainfall downwind.

The goal for 2020 is to stop the expansion of fossil fuel use, which some call "stabilizing" emissions. That word sounds good, until you discover how minimal it is. "Stabilizing the patient" is what you do in the emergency room to keep the patient from crashing for good; it only buys time for more definitive treatment in the operating room. Even if we stop the growth, we'd still be adding a constant amount of fossil carbon to the atmosphere each year. Stopping that before 2040 would be my second goal. Cleaning up the accumulated excess by 2080 would restore the CO_2 concentrations to what they were in 1939 and reverse many of the climate changes.

If we don't start thinking big about the CO_2 problem, we may miss our opportunity to stop a climate runaway that will trash the habitable parts of the earth. We used to be able to say, "If you can't stand the heat, get out of the kitchen." But it's not going to be practical to get off our only habitable planet.

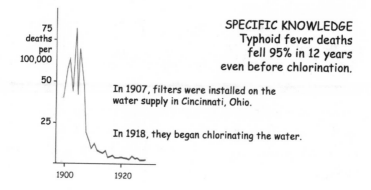

SPECIFIC KNOWLEDGE
Typhoid fever deaths
fell 95% in 12 years
even before chlorination.

In 1907, filters were installed on the water supply in Cincinnati, Ohio.

In 1918, they began chlorinating the water.

Since there is a lot of climate science left to be discovered, one source for optimism comes from the analogy to how things went in medical science when it was in a comparable stage of development. A lot can be done with inexpensive technology once you understand how things work.

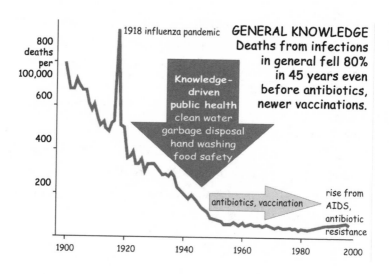

1918 influenza pandemic

GENERAL KNOWLEDGE
Deaths from infections
in general fell 80%
in 45 years even
before antibiotics,
newer vaccinations.

Knowledge-driven public health clean water garbage disposal hand washing food safety

antibiotics, vaccination

rise from AIDS, antibiotic resistance

18

The Climate Optimist

Mention global warming at a social gathering and see what happens, now that skepticism and glib comebacks have turned into concern and sorrow. People will, of course, assume that you're a pessimist about our prospects.

"Not really," I protest. That earns me a quizzical look.

"Wait a minute," she says. "If you're an optimist, why do you look so worried?"

Dramatic pause.

"So you think it's easy, being an optimist?"

Indeed, many scientists look worried these days. We've had a steady diet of bad news coming from climate scientists and biologists.

But I think that, with notable exceptions, we scientists are more hopeful about weathering this crisis than are people who only recently have become aware of how serious the situation has become—but cannot yet imagine a way out. To become even a guarded optimist, you have to think hard about both a technofix and some social engineering.

First, I reflected, the history of science and medicine shows that, once you mechanistically understand what's what, you can approach all sorts of seemingly unsolvable problems. I'm optimistic that we will learn how to stabilize climate.

When pessimism tempts me, I usually remember the progress that I've seen. When I was born in 1939, antibiotics were just a rumor, there were few vaccines, and one's chances of dying from an infection were three times greater than they are now.

I've seen an enormous increase in our knowledge about how bodies work, from molecules to mind (what I'd still be working on, were it not for the climate crisis). The average lifespan has been extended by decades in many countries, just in the time that I've been personally observing the scene as a medical school professor. And in the first half of the twentieth century (graph, page 226), deaths from infections dropped by an order of magnitude even before antibiotics and vaccinations came to dominate the scene.

Just the basic knowledge about how diseases spread was what did most of the job, not a needle. Once this new knowledge was incorporated into everyday hygiene practices, eight out of ten fatal infections were prevented. We're used to thinking of science discoveries leading to technological innovation. But here you see how knowledge, pure knowledge, pays off all by itself.

The reason I'm not so pessimistic about climate is that, once you understand what's what, progress speeds up. It is reasonable to hope that we will learn how to intervene and restore the climate in the decades ahead, much as we earlier did as basic knowledge transformed the practice of medicine.

But the problem is clearly broader than scientists delivering new knowledge. Unless we redesign our civilization in numerous ways, all of the science in the world won't save us. Politicians and citizens alike have been deaf to scientific warnings over the last half century, reluctant to spend on infrastructure and education, and more concerned with present profits than with their responsibility to future generations.

Unfortunately the window of opportunity, to act on such knowledge, is closing. Fifty years have now passed since the first unequivocal scientific warnings of an insulating blanket of CO_2 forming around the planet. We have already entered the period of consequences. Climate scientists have long been worried about their children's future. Now they are also worried about their own.

Our Faustian bargain over fossil fuels has come due. Goethe's Dr. Faustus had twenty-four years of party now, pay later—and indeed, it's been longer than that since President Ronald Reagan axed the U.S. budget for exploring alternative fuels. This led to doubling our use of cheap coal, the worst of the fossil fuels. The energy

companies are planning, under business-as-usual, to
redouble coal burning by 2030—even though we can now
see the high cost of low price.

The devil's helpers may not have come to take us away,
but killer heat waves have started, along with extreme
weather that keeps trashing the place. We're already
seeing droughts that just won't quit. Deserts keep expanding. Oceans keep acidifying. Greenland keeps melting.
Dwindling resources are triggering terrible genocidal wars
with neighbors. All of them will get worse before they get
better.

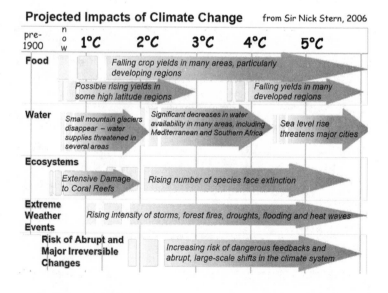

Projected Impacts of Climate Change   from Sir Nick Stern, 2006

Worse, a tipping point can lead to an irreversible
demolition derby. Should another big El Niño occur and
last twice as long as in 1983 or 1998, the profound drought
could burn down the rain forests in Southeast Asia and the

Amazon—and half of all species could go extinct as a result.

Carbon-free energy is something that we simply have to do. The time for talk is past. If we turn around net carbon emissions by 2020 rather than 2040, we get another 2° of fever rather than 3°—and that's a big difference.

Remember, I tell her, that another $2^{\circ C}$ is already going to mean substantial sea-level rise from melting Greenland—and even $1^{\circ C}$ in the tropics will reduce crop yields for the cereal grains.

"I see why you're worried," she says, getting a word in edgewise. "But what's your optimistic scenario for dealing with this fossil-fuel fiasco?"

For starters, I think it likely that the leaders of the major religious groups will soon come to see climate change as a serious failure of stewardship.

And once they see our present fossil-fuel use as a deeply immoral imposition on other people and unborn generations, their arguments will trump the talk-endlessly-to-buy-time objections of the vested interests—just as their moral arguments did when ending slavery in the nineteenth century, in spite of the perceived costs to the economy.

Second, the developed nations are fully capable of kick-starting our response to global fever with present technology—enough to achieve, within ten years, a substantial reduction in their own fossil fuel uses. How?

Wind farmers will prosper as pastures grow modern windmills to keep the cows company. Heat farmers will drill down deep. Giant parking lots, already denuded of trees, are perfect places for acres of solar paneling. Drivers will love the shaded parking spaces. Maybe they'll even re-charge their cars.

The escalating Carbon Fee ("C-fee") with an elimination of payroll taxes will create a big wave of retrofitting homes and businesses. Value-added C-fees will give every stage in the supply chain an incentive to go C-free. Just call it "C-free or C-fee."

Big, brightly lit grocery stores with giant parking lots will compete poorly with efficient warehouses that deliver web and phone orders within the hour, like pizza. Smaller neighborhood grocery stores will once again do a big walk-in business and they will compete with the ware-houses by offering "green bicycle" delivery powered by excess adolescent energy.

High-speed toll gates will become the norm on com-muter highways. (Yes, I know, but remember that the paycheck was just enriched by eliminating payroll taxes.) Since 90 percent of U.S. commuters go by car, and only one of nine drivers has a passenger, splitting the cost will become attractive.

Speed limits will be lowered to 50 mph (80 km/h) for fuel efficiency and, as in 1973, drivers will marvel at how smoothly the traffic flows when everyone is going the same speed. Double taxes will apply to vehicles with worse-than-average fossil-fuel consumption, steadily reducing the number of oversized vehicles with poor streamlining. Plug-in hybrids will begin to dominate new car sales. In an era of more power grid failures, hybrids will serve as emergency generators for the house. All-electric "clean cars," while they are parked for a charge, will help prevent grid failures by smoothing dips and spikes.

A firm, fast schedule will be established for retiring or retrofitting existing coal plants. My bet is that adding geothermal or nuclear power plants will prove safer, cheaper, and faster than fixing coal.

So which numbers are the most important to keep in mind? Emissions each year, emissions growth, CO_2 concentration, or temperature?

It depends. For ocean acidification, the CO_2 concentration in surface waters is the big player in the long-run. But the heat is the big short-term problem.

For climate change, temperature change per se is the driver that precipitates many other consequences. For producing temperature change, it's not only CO_2 that drives it but the other greenhouse gases, such as methane from pipeline leaks and nitrous oxide from fertilizers.

Remember that decreasing sinks via land clearing can be just as important as increasing sources. It acts like an increased carbon emission.

One of the first things that you notice from the emissions pie (back on page 158) is that land clearing accounts for 18 percent of the problem, even larger than the worldwide use of oil for transportation at 14 percent. And that agriculture adds 14 percent too, mostly via fertilizers, plowing practices that speed soil decomposition, and cows that burp. So shrinking this carbon pie down to half its present size depends heavily on agricultural innovations, as well as overhauling the present energy production portfolio. While two-thirds of the pie comes from the urban energy emissions that we usually talk about, a third of the possibilities for shrinkage come from the countryside's excesses.

So an optimist notes that fully a third of our possibilities for reversing the trends come from a sector that we don't usually talk about. And that reducing methane emissions, as we have already started doing, is a particularly fast way of reducing the fever.

Before 2020, let us assume that the transition to hybrid electric and compressed-air vehicles will shift transportation's energy demands to C-free but industrial-strength sources such as hot rock and nuclear. They will power much of the transportation sector, thus reducing oil use.

The low-loss DC transmission lines should allow, via cables under the Bering Strait, solar-generated electricity to flow from the bright side to the dark side of the earth, around the clock. Superconducting wires will likely be running inside some retired pipelines.

By 2020, we ought to see some important new technology coming on line, not just improvements in what we already use. The highly efficient binding energy extractors (BEEs, the fourth-generation nuclear power plants) could be running on the spent fuel of the earlier generations.

We need a way of pumping down the excess atmospheric CO_2 that is cooking the earth, in the manner that a kidney dialysis machine cleans the blood. Perhaps we will succeed in enhancing existing sinks. For example, we might encourage the whale's favorite food, the tiny plankton which provide half of the oxygen we breathe as they separate the C from the CO_2.

Sometime before 2040, I'd bet that we will be mining the air for CO_2, using it as fuel via some artificial process analogous to photosynthesis. But even if we invent and debug such schemes tomorrow, it can take several decades before an invention makes a dent in our increasingly urgent problem.

By 2040, let us suppose that we are busy extracting more CO_2 from the atmosphere than we add.

This will only happen if the technology of the developed world has become good enough to compensate for what's still going on in the overstressed nations that are too disorganized to get their energy act together.

When CO_2 levels fall enough to counter the delayed warming from past excesses, we will begin to see a reversal of droughts and heavy weather, though the rise in sea level will likely continue, a reminder to future generations of our twentieth-century Faustian bargain. Even if Greenland and Antarctica stay frozen in the summertime, the distressed foundations of the sides of the Greenland ice sheet may allow the ice dome to spread sideways and push ice into the ocean, long before it might melt.

In 2006, climate scientist Jim Hansen said that "[W]e have at most ten years—not ten years to decide upon action, but ten years to alter fundamentally the trajectory of global greenhouse emissions . . . we are near a tipping point, a point of no return, beyond which the built in momentum and feedbacks will carry us to levels of climate change with staggering consequences for humanity and all of the residents of this planet."

We need to turn on a dime, close to what we saw in the United States in 1940 in response to the ominous Nazi takeover of Europe. President Franklin D. Roosevelt asked Congress to increase the construction of military aircraft by ten-fold and was wildly cheered. "The President's big round number was a psychological target to lift sights and accustom planners in military and industrial circles alike to thinking big," wrote a military historian. Roosevelt used the metaphor of a "four alarm fire up the street" that needed to be extinguished immediately, whatever the cost.

From a standing start in late 1941, the automakers converted—in a matter of months, not years—more than 1,000 automobile plants across thirty-one states . . . In one year, General Motors developed, tooled, and completely built from scratch 1,000 Avenger and 1,000 Wildcat aircraft . . . GM also produced the amphibious 'duck'—a watertight steel hull enclosing a GM six-wheel, 2.5 ton truck that was adaptable to land or water. GM's duck 'was designed, tested, built, and off the line in ninety days' . . . Ford turned out one B-24 [bomber] every 63 minutes.

—author Jack Doyle, 2000

Now there's a source of optimism: we did it before. With great challenges come great opportunities and I'm an optimist about our ability to respond with innovation, much as various countries did during World War II.

Most progress against air pollution has been cheaper than expected. Smog controls on automobiles, for example, were predicted to cost thousands of dollars for each vehicle. Today's new cars emit less than 2 percent as much smog-forming pollution as the cars of 1970, and the cars are still as affordable today as they were then. Acid-rain control has cost about 10 percent of what was predicted in 1990, when Congress enacted new rules. At that time, opponents said the regulations would cause a "clean-air recession"; instead, the economy boomed...

Americans love challenges, and preventing artificial climate change is just the sort of technological and economic challenge at which this nation excels. It only remains for the right politician to recast the challenge in practical, optimistic tones ... But cheap and fast improvement is not a pipe dream; it is the pattern of previous efforts against air pollution. The only reason runaway global warming seems unstoppable is that we have not yet tried to stop it.

—commentator Gregg Easterbrook, 2006

It is becoming clear that we must make a choice. We can resolve to move rapidly to the next phase of the industrial revolution, and in so doing help restore wonders of the natural world, of creation, while maintaining and expanding benefits of advanced technology.

Or we can continue to ignore the problem, creating a different planet, with eventual chaos for much of humanity as well as the other creatures on the planet.

—climate scientist Jim Hansen, 2007

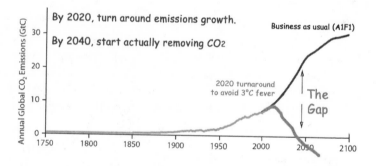

This graph shows the slow growth in fossil-fuel emissions until about 1950, then the five-fold increase in the second half of the twentieth century, and finally the projected path until 2100 under a Business As Usual economic scenario.

The gray line is simply my cocktail-napkin sketch of what is needed to reverse the growth in emissions by 2020 and getting into net removal of CO_2 by 2040. The vertical distance between the two projections, which must be filled by reducing emissions and creating carbon sinks, is called The Gap.

19

Turning Around by 2020

The future ain't what it used to be.
— Yogi Berra, Baseball Hall of Fame philosopher

Time has become so short that we must turn around the annual emissions growth before 2020 to avoid saddling today's students with the world of refugees and genocides that results if we're too slow.

That means not waiting for a better deal on some post-Kyoto treaty. It means immediately scaling up technologies that we know will work, not waiting for something better that could take decades to debug.

The standard green answers (compost, carpool, eat locally grown foods rather than ones that require long-haul transport, become a vegetarian, and so forth) are all important. But, as James Lovelock likes to say, they may prove no more effective than dieting.

What we need are sure-fire solutions that stop the CO_2 pollution from all of those tailpipes and smokestacks. And do it quickly, which means not relying on efficiency improvements or new rapid transit systems that take decades to implement. Our problem has now become too big and too immediate to rely on reforming people's habits.

Since a quick response is not the timescale of investment capital, the energy changeover is going to require major government leadership to make sure it gets done quickly. The Manhattan Project of 1942 to 1945 shows us how we have quickly turned recent science discoveries into major engineering projects.

Going to the Moon was a major national effort that, while expensive, did not require a wartime economic restructuring. I had lunch with George Mueller, who ran the Apollo Project for NASA in those critical years from 1963 to 1969. I asked him what it would take to stage, on an urgent basis, our energy makeover and climate restoration.

First, he said, simply banning certain energy uses would not work any better than the U.S. experiment with banning alcohol, which simply created a bootlegging industry. (Imagine cheap Chinese incandescent bulbs smuggled into California, Australia and Canada, now that they have decided to ban the old-fashioned bulbs.)

For an Apollo-scale project to create non-carbon energy alternatives, Mueller said that we needed a goal that was easy to understand (something like putting a man on the moon and returning him safely). And the goal needed a time frame (President Kennedy's "this decade") to persuade the public to act now.

> We choose to go to the moon
> in this decade
> and do the other things,
> not because they are easy,
> but because they are hard,
> because that goal will serve
> to organize and measure
> the best of our energies and skills,
> because the challenge is
> one that we are willing to accept,
> one we are unwilling to postpone,
> and one we intend to win . . .
> —John F. Kennedy, 1962

I'd propose *Turning around by 2020* as our goal and time frame, followed up by two more.

1. The 2020 target would be stopping the annual growth in emissions to keep the eventual fever below 2°C. But we'd still be growing the CO_2 blanket, year after year, just at a constant rate.

2. The 2040 target would be to stop the annual CO_2 growth altogether. This means that increased sinks would have to balance out any remaining fossil carbon emissions, including the delayed ones. Note that we still haven't reduced CO_2 concentrations, only stopped its growth. Then we begin removing more CO_2 from the air each year than we add. That makes it a matter of adding sinks, not merely controlling emission sources. Call it *Sinking CO_2 by 2040*.

3. Restoring CO_2 concentrations to 1939 levels would be my third goal. Call it *Restoration by 2080.*

We may well need to double power production in order to clean up CO_2, double again for worldwide modernization, and with another step up if we are to go to electric cars. This expansion only makes sense with C-free energy—lots of it.

Let me now evaluate the various candidates for accomplishing the 2020 turnaround. I'll later summarize their advantages in a table for easier comparison.

Use less. That's the most obvious solution. There are two versions. The relatively painless one is increasing efficiency. A modern refrigerator uses one-fourth the power of a 1975 model. And so we replace incandescent light bulbs with compact fluorescents or LEDs. Same lumens, less watts. Better gasoline mileage and more carpools also achieve the same end use, but using less power.

The more painful version is the diet that requires shedding the end uses themselves—say, turning off the all-night street lights or hanging the clothes out to dry instead of using the clothes dryer. Such banished end uses tend to creep back on stage within a few years.

Furthermore, both versions are local or regional solutions that won't produce global solutions in time.

Developing countries won't forego modernization just because we say so.

Better for 2020 to assume that those end uses will stay the same and even expand. And so we must focus on substituting C-free power sources and finding ways to create new carbon sinks.

The lights left on, all night long.

You already know the scene for reducing our use of oil—converting to hybrid vehicles and alternative fuels. If plug-in hybrid electric vehicles (PHEVs) were to replace the 198 million cars, vans, SUVs, and light trucks in the U.S., it would cut oil imports by half.

Though I wouldn't recommend it, 84 percent of the recharging job could be done with excess overnight capacity in America's coal-fired plants. Even though burning coal to replace oil, it would nonetheless reduce overall CO_2

emissions. Such is the wasted energy from using 198 million inefficient internal combustion engines that must be kept idling in congested traffic.

But this has to be done globally. A fleet of PHEVs requires much mining to make the batteries. Poorer countries would have to import them. "Air cars" that run on compressed air would be easier for a developing country to manufacture from local materials.

No, it's not a rocket. No electric motor, either. It's an engine where the compressed air runs a piston. Refilling the air tank can be done overnight by plugging in the on-board compressor. So air cars also run on electricity, just one step removed. It's the same for hydrogen fuel cell cars.

Just as a spray can cools your hand, so the carbon-fiber air tanks will become quite cold during use. The free air conditioning ought to make air cars popular in the tropics—and elsewhere, as global warming increases. I predict that beer will be cooled this way and that attached garages will also become popular, opening widely into the living quarters to cool them down as you unload the groceries.

	Ability to expand	Public view	Down side	Ups & Downs	Enough by 2020?
Hybrids	large	very good	mining battery	—	●●●
Compressed air car	large	none yet	air tanks	—	●
Improve efficiency	good	in favor	slow grind	—	●
Dieting	limited	a pain	easy to fail	yo-yo	no

oal-fired power plants are the big actors on the fossil carbon scene (in the U.S., more than half of the electricity comes from burning dirty coal) and if we don't address them immediately, the long run considerations will be irrelevant.

Coal-fired power plants are large, what with the ash heaps and settling ponds for the parts of coal that aren't carbon. Their footprint also includes those sawed-off mountains and terraformed landscapes left behind.

Though some new coal plants do a better job of capturing the metals which fall out locally, and the sulfuric acid that can travel much farther downwind, capturing the invisible CO_2 and methane is talk rather than action. The so-called "clean coal" (regularly featured in the quarter-page greenwashing ads on the *New York Times* op-ed page) is, at present, just trapping more ash and sulfur from going up the smokestack.

Coal-fired power plants throughout the world are the major source of radioactive materials released to the environment. Thorium and uranium may only be a tiny fraction of the coal but we burn a lot of coal. These trace amounts add up to far more than the entire U.S. consumption of nuclear fuels for electricity. About 10 percent is carried aloft on fly ash, made airborne for us to breathe.

For all the talk of capturing CO_2 and storing it down deep somewhere, it looks like such technology will suck up 40 percent of the power generated. Let's see, retrofitting the 403 existing U.S. coal plants would create a need for another 269 coal plants. Big Coal's sales would go up 67 percent. (It's odd that no one ever mentions that.)

two steam turbines

steam pipes

750 mw generator

starter motor

The clean side of a 750 megawatt power plant, excluding the source of the steam (coal, in this case).
It's almost big enough to power the City of Seattle.   (c) George Campbell 1993

The Zimmer power plant in Ohio was supposed to be a nuclear power station but, in the middle of construction, they switched to coal, abandoned the expensive containment dome next door, and now truck the ash to what will become the highest hill in southern Ohio. Dumping the finer stuff into the air we breathe is "free."

Over a fifty-year lifetime, each retrofitted 500-megawatt coal plant would produce a billion barrels of liquid CO_2 to be stored underground. No one knows how safe such storage would be. An earthquake could fracture the well's casing, allowing the stored CO_2 to escape. Clearly, this is experimental technology, not ready for prime time.

Such capture-and-storage talk may be another example of Big Coal trying to buy time by delaying action while, of course, getting yet another tax break from Congress to increase their record profits. Worst of all, even if practical, carbon capture and storage is not going to help very much for decades. I can think of better ways to spend our climate makeover money.

In 1992, Zimmer set the world record for the most coal burned by a single generating unit, consuming four million tons of coal that year and venting thirteen million tons of CO_2.

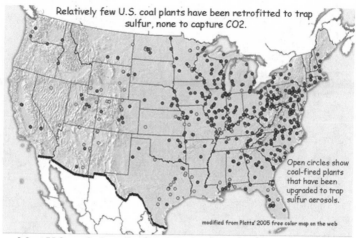

Relatively few U.S. coal plants have been retrofitted to trap sulfur, none to capture CO2.

Open circles show coal-fired plants that have been upgraded to trap sulfur aerosols.

modified from Platts' 2005 free color map on the web

Most U.S. coal plants have not bothered to retrofit acid rain scrubbers. Even using low-sulfur coal, there is the problem of uranium and thorium in the fallout.

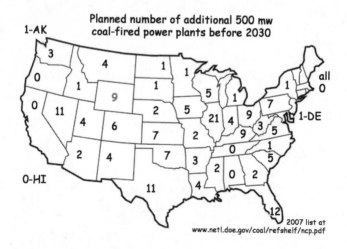

Planned number of additional 500 mw coal-fired power plants before 2030

2007 list at www.netl.doe.gov/coal/refshelf/ncp.pdf

It's grossly irresponsible, but U.S. power companies are planning on doubling coal consumption by 2030. The U.S. Department of Energy 2007 report is entitled "Coal's Resurgence in Electric Power Generation." It contains not a single word about the impact on climate change.

Here's my subjective evaluation of coal.

	Ability to expand	Public view	Down side	Ups & Downs	Foot print	Storage needed	Enough by 2020?
Coal as usual	huge	dirty	huge	no	very large	no	n/a
Coal but capture the CO2	large	caution	storage burp, leaks	no	67% more coal	huge	very little

H ot Rock Energy is the most attractive possibility I know for quickly expanding an alternative energy source. Drill a 5-km-deep well into hot granite, feed it water, harvest the rising steam to spin the usual old-technology steam turbine, and you get electricity.

Though still in the demonstration project stage, it doesn't suffer from nuclear fears, 15-year permit delays, and 5-year construction times. It is an alternative energy solution that is C-free, doesn't fade as the sun sets, isn't fickle like the wind, doesn't require lots of space like biofuels, and doesn't require mining heavy metals that are radioactive. It's nice and steady without needing storage like hydro. It's immune to droughts.

There's nothing equivalent to coal trains and super-tankers, not even trade deficits. There are no big questions hanging over it as with carbon capture and sequestration. A Hot Rock plant's footprint is no bigger than a two-story parking garage, with no runoff or air pollution or trucks hauling stuff—indeed, it would fit atop an old oil platform offshore or inside a large barn (except when cooling condensers are needed).

If you haven't heard of Hot Rock Geothermal (and it is typically left off the alternative fuels list, even when *Scientific American* did a special issue on the subject in 2006), it's because "geothermal" has an image problem rather like electric cars once had. It took the success of a 1997 gasoline-electric hybrid called the Prius to help people think ahead to an all-electric car without defaulting to an image of a golf cart of limited utility, not suitable for the freeways.

Hearing geothermal, we often pop up a mental image of a sulfurous hot spring and wrinkle our nose. Too many people think that geothermal is just piping near-surface hot water around to heat some buildings—say, Idaho's State Capitol buildings in Boise. This in turn makes you think that geothermal electrical power is a special case, nice for Iceland but not more generally. That, however, is your grandfather's notion of geothermal.

And a heat pump might be your father's. That's a different principle, the one that has long led people to build underground cellars to store food. A few feet down, the soil doesn't change temperature very much between winter and summer—and so by running a pipe through it, a heat pump can get it to cool water (which then cools air) in the summer and to provide some heat in the winter. Just think of burying a sprinkler system without the sprinklers.

Many countries have good traditional geothermal resources that have yet to be exploited for generating electricity. Shallow wells are the most common. That's the "geothermal" implied in most mentions of alternative fuels.

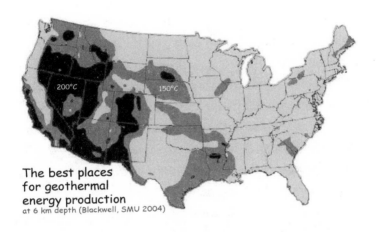

The best places for geothermal energy production
at 6 km depth (Blackwell, SMU 2004)

The deep version now coming on stage is Hot Rock Energy with two or three adjacent wells. The idea is *not* to find hot water but rather hot rock that is dry. Then apply water to make steam. Though the U.S. has been lagging behind, the Hot Dry Rock concept was invented by scientists working at the Los Alamos National Laboratory in 1972.

Below the sedimentary layers is usually granite that's hot and dry. The farther down you drill into granite, the hotter it gets. Drilling 6 km below the surface is often sufficient to get 200°C (about 400°F, oven temperature) in the western U.S. The 100°C you'd get elsewhere works too, though it produces lesser amounts of electricity. (So you drill twice as many wells.) It usually takes the deep drilling technique used by the oil industry which can go 7 km down.

Unlike the hot springs version of geothermal, you have to provide your own water. But after you prime the well, you just re-circulate. What comes up as dry steam is pumped right back down again as water, via a second well nearby. It forces through cracks in the granite, heats back up, flashes into steam, shoots up the other well to the steam turbine, which spins the electrical generator, which feeds the great electrical grid, which keeps your domestic climate comfortable and your car recharged.

And how do these two wells connect? Such deep rock is already fractured along onion-like sheets, ancient planes of stress from bending. Mineralization has filled those cracks—but high-pressure injection can force water into them, opening up passages. When the high pressure is released, many do not reseal. Sometimes the layers shift a little, and the noise from such little earthquakes serves to locate the newly-opened crack. A map of the enhanced fracture zone is built up and, when it is several km across, the second (and sometimes a third) well is drilled into it to harvest the steam.

Gushers and mud eruptions don't come up out of the granite layers. If a sizeable earthquake fractures the well shaft, nothing happens—you just drill a new well nearby. That makes it much safer than drilling for oil or natural gas—or for storing CO_2.

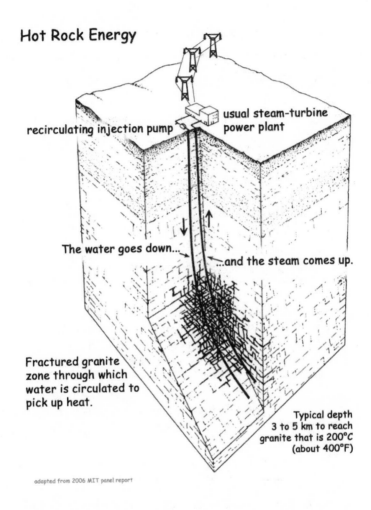

Hot Rock Energy

recirculating injection pump

usual steam-turbine power plant

The water goes down...

...and the steam comes up.

Fractured granite zone through which water is circulated to pick up heat.

Typical depth 3 to 5 km to reach granite that is 200°C (about 400°F)

adapted from 2006 MIT panel report

Hot Fractured Rock is drought-proof and does not involve a perpetual stream of truck traffic as biofuels and fossil fuels do. It is perhaps the least demanding on industry, except for manufacturing enough tall drill rigs and training enough crews. What's above ground is mostly modern steam plant gear, manufactured in many countries and quite reliable. Operating it is well within the competence of all developing countries.

How extensive a resource is deep geothermal? For the U.S., the experts said it could yield a thousand times more than our present overall energy use. How polluting? Close to zero.

> Geothermal energy from EGS [enhanced geothermal systems = Hot Rock Energy with engineered fracturing] represents a large, indigenous resource that can provide base-load electric power and heat at a level that can have a major impact on the United States, while incurring minimal environmental impacts. With a reasonable investment in R&D, EGS could provide 100 GWe or more of cost-competitive generating capacity in the next fifty years. Further, EGS provides a secure source of power for the long term that would help protect America against economic instabilities resulting from fuel price fluctuations or supply disruptions. Most of the key technical requirements to make EGS work economically over a wide area of the country are in effect, with remaining goals easily within reach.
> —The MIT panel's 2006 report, entitled "The Future of Geothermal Energy: Impact of Enhanced Geothermal Systems (EGS) on the United States in the 21st Century."

Hot Rock Energy, unfortunately, has been on the back burner for decades, along with most other alternative energy sources, kept there by cheap-and-dirty coal and the small budgets for government R&D. Nonetheless there have been various research projects around the world that have demonstrated the deep heat mining techniques over the last three decades.

Serious power production, however, is only getting started. In northern France, they are getting near-commercial-sized yields at depths of 4 to 5 km. There are some projects in southern Germany, northern Switzerland, and Japan. Australia has quite a few proof-of-concept projects limping along on private money.

125 megawatt two-well geothermal power station in the Philippines

A modern two-well geothermal plant, though using shallow wells and a heat exchanger (thus requiring 178 condensers) not needed for dry steam. Nothing is more than two stories high. It was operating 15 months after the ground-breaking ceremony in Lyete.

The only hesitation that I have about Hot Rock Energy for 2020 is that there is simply not enough experience with it yet, compared with the experience of running hundreds of nuclear plants over fifty years time. Even though merely combining two tested techniques, steam power plant and deep drilling/stimulation, there will be beginners' errors to discover.

The capital costs per megawatt-hour are similar to those of a new coal plant. They are mostly drilling costs. Indeed, until opening up those fractured rocks in the depths with

the initial high-pressure injection, you don't know what
size power plant to order for the well head. That might
cause private capital to hesitate, suggesting a proper role
for government money to do the initial steam farms.

As demand increases, improvements will likely drive
down drilling costs. Hot Rock power plants could be
rather simple compared to shallow geothermal plants
today, where the well's output contains a lot of things that
you wouldn't want to inflict on a steam turbine sensitive
to corrosion. Protecting it means a heat exchanger and that
requires a hundred condensers to cool the secondary fluid
before it re-circulates through the steam turbine.

So a lot of customizing attends most geothermal today.
But continuing further down to 150°C dry granite would
allow mass production techniques for simplified power
plants. Each installation can tie up a deep drilling rig for
the better part of a year, so we are going to need to clone
those tall rigs.

If I were the 2020 czar, I'd place an order for twenty
deep drilling rigs and fund fifty small heat farms in order
to find the beginner's errors and the efficient combi-
nations. We urgently need to know if Hot Rock Energy can
be ramped up worldwide to thousands of units.

	Ability to expand	Public view	Down side	Ups & Downs	Foot print	Storage needed	Enough by 2020?
Hot Rock Energy	huge	Just another well?	Month of small EQs?	very stable	very small	none	●●●● to ●●

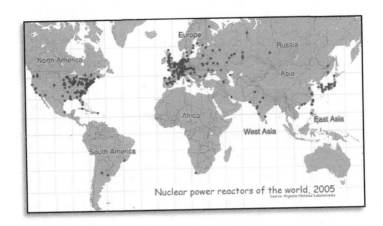

Nuclear power reactors of the world, 2005
Source: Argonne National Laboratories

Nuclear power generation is currently the major C-free energy source. It is over fifty years old, with an excellent safety record. It took three decades before the efficiency doubled. Unlike the other expandable C-free sources, the beginner's mistakes have already been made.

As I mentioned, France has nearly quadrupled electricity production using 78 percent nuclear. It sounds as if nuclear power is cheap in such quantities. So much for arguments that nuclear power is expensive and that reform of our dependence on fossil fuels is impractical, can't be done, will damage the economy, and other excuses heard for maintaining the status quo (and current fossil-fuel profits). Why are many so countries denying themselves this C-free power source, while allowing growth in the hazardous fossil fuels?

Let me briefly discuss the downside of nuclear power. Our view of it—including my own view, until recently— tends to focus on fuel diversion into nuclear weapons or dirty bombs, reactor accidents, and the long-term management of nuclear waste.

Some things have changed since the heyday of the anti-nuclear-power movement in the 1970s (though not Ralph Nader). Since the Soviet Union's political meltdown, the cat may already be out of the bag for illicit nuclear fuel, so that avoiding additional reactors may not gain us much.

New issues have also emerged. There is now the problem of suicide aircraft attacks with a full fuel load, which might scatter radioactivity downwind. There may be undergrounding solutions to this if the containment walls cannot be strengthened enough, but again pause and note that additional reactors do not really increase this problem; there are sufficient targets already. Indeed, chemical plants of many types are vulnerable. In Bhopal, India, all it took was a gas leak at a pesticide plant to kill 8,000, injure a half-million people, and contaminate an entire city.

There is much data on the safety and environmental problems of all the power sources. Nuclear electricity generation has proven far safer than fossil fuels of all sorts, and even safer than hydro. Dams fail. Per megawatt generated, the hydro fatality rate around the world is a hundred times higher than for nuclear electricity.

That is startling enough. The production and storage of the fossil fuels is far more deadly. Counting only the major disasters (each big enough to kill 300 or more) between 1979 and 2006, there were more than 2,400 deaths from oil and 1,800 from natural gas. For coal—well, China alone has 6,000 miner deaths *each year*. Coal mining in the Ukraine is even more deadly.

The worldwide fatalities from nuclear power generation average out to one per year. For commercial nuclear power in all countries except one, there has never been a fatality. (An experimental military reactor accident killed three operators in 1961. There have also been on-site fatalities from bursting steam pipes not directly connected to the reactor, a problem with old steam pipes in general and one that high-tech inspection techniques ought to eliminate.)

Two workers at a small, badly designed nuclear reprocessing plant in Japan were killed in 1999 in a flash of radioactivity. No radiation was released into the environment. It was not at a reactor site. The plant was a small specialty operation, not part of the commercial fuel cycle for electricity. They were processing a batch of fuel that had been enriched about four times more than allowed in any commercial nuclear power plant and—the fatal error—they didn't dilute it properly.

The Chernobyl reactor meltdown in 1986 is the only major meltdown accident with fatalities. The operators had violated the rules by disabling major safety features and, when the power surged and popped the lid, it had—incredible as it now seems—no containment to trap the radioactive gases. In the first few weeks, thirty-two Ukrainian staff and firefighters died.

Immediate fatalities are the only number for which easy comparisons can be made between energy sources. So that's less than fifty killed in the first fifty years of nuclear power reactors, all in one accident.

What about Three Mile Island in Pennsylvania? This accident occurred in 1979 (just thirteen days after a star-laden movie opened, a nuclear reactor control room drama, *The China Syndrome*). The steam explosion killed no one. (No injuries, either.) Though it was a cliffhanger, the release of radioactivity was largely confined to the site (unlike Chernobyl, it had a good containment vessel with absurdly thick walls). Anyone living nearby got a one-time dose less than what they got every day from the rocks beneath their house.

Delayed deaths are often difficult to attribute to a single cause, making comparisons between power plant types even more problematic. But for Chernobyl, we can safely say that another twenty-five died later, including nine children from thyroid cancer, but that the feared spike in leukemia did not materialize. Guessing farther out into delayed effects, perhaps 1 percent of the 200,000 workers exposed in the accident and during its cleanup may die from their radiation exposure, suggesting that 2,000 eventual deaths from the accident are possible.

Of course, mining coal has similar delayed effects, such as black lung disease. They involve many more people, including non-miners who simply live downwind of coal-burning power plants (in the U.S., that's almost everyone in the eastern half of the country) and breathe the ash and sulfur aerosols. Similarly, petroleum causes many delayed deaths from air pollution.

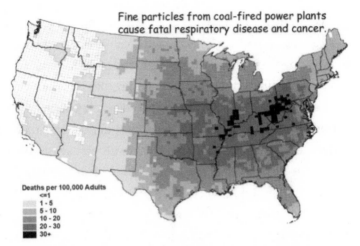

Fine particles from coal-fired power plants cause fatal respiratory disease and cancer.

Deaths per 100,000 Adults
<=1
1 - 5
5 - 10
10 - 20
20 - 30
30+

Three Mile Island had a huge impact on the future of nuclear power in the U.S. and in other countries, creating a gap that has been largely filled by coal. No new nuclear plants have been started in the U.S. since 1978, though new nuclear plants are common enough in the rest of the world (thirty-one countries now generate nuclear power).

We already know how to make safe nuclear reactors, even for the traditional style that uses water to both cool the reactor and to slow down the neutrons so they don't trigger additional, unwanted nuclear fissions. The danger here is that, if the water leaks out or the pumps fail or the water boils off, the reaction speeds up and heats up. And so you get a meltdown of the core and a radioactive slag heap in the basement.

There is not an explosion as with a nuclear bomb. These are steam explosions, the same as when the lid of a press-ure cooker pops and coats the kitchen with hot food. The reactor may also catch on fire. Any steam or ash that escapes from an overheated reactor can create radioactive fallout downwind (hence the containment dome, what kept Three Mile Island from being a problem offsite).

There is now, fifty years after the first nuclear power station was built, a much safer third-generation reactor design that uses a water tower above the reactor. Water floods the reactor if it overheats, all without relying on pumps or operator actions.

Given that we need something sure-fire, we have no choice but to start expanding nuclear. Clearly, nuclear is capable of being a big part of the solution but there are doubts in my mind about whether permit obstacles will be hurdled in time.

With any luck, we'll be able to cancel one nuclear order after another with scale-up successes in alternative C-free fuels. But order we must.

For the long run of 2025, there's a design for a fourth generation reactor that doesn't rely on water for slowing down the neutrons. Like a fast-breeder reactor, it runs on fast neutrons and thus generates all manner of radioactive isotopes. It extracts twenty-four times as much energy from its fuel pellets as conventional reactors do. This leaves the fuel exhausted and unsuitable for bomb manufacture, licit or illicit. That may handle the traditional worries about fuel diversion into nuclear weapons, what we saw in 1974 when India illicitly made a bomb using a research reactor donated by Canada, with heavy water supplied by the U.S.

Furthermore, the fourth-generation nuclear waste decays to ordinary levels in only centuries, not the 10,000-year timescale of the current nuclear waste which has only had 4 percent of its binding energy extracted. The isotopes with the long half-lives are broken up by the fast neutrons. So the timescale for managing stored nuclear waste shrinks by 96 percent.

In principle, we know how to solve the recycling aspect as well. Because the fourth generation is so much more efficient at extracting megawatts from uranium, they can run on the accumulated "spent" fuel of the last fifty years, solving our storage dilemma for high-level nuclear waste. And when the fourth-generation fuel's output drops off because of accumulating lighter elements that soak up neutrons, the fuel pebbles can be reprocessed on site rather than being shipped long distances (South Africa, for example, ships its spent fuel to France; oddly, U.S. commercial nuclear plants are not allowed to reprocess fuel, period.)

As the authors of a *Scientific American* article in December 2005 write, this fourth-generation design "could overcome the principal drawbacks of current methods—namely, worries about reactor accidents, the potential for diversion of nuclear fuel into highly destructive weapons, the management of dangerous, long-lived radioactive waste, and the depletion of global reserves of economically available uranium."

Much of our traditional rationale for opposing expansion of nuclear power (or even, as Germany plans to do, retiring all nuclear power plants) needs reevaluating. One of the great hurdles is the public's perennial confusion of nuclear electricity generation with nuclear bombs.

I have a suggestion: let us rename the fourth-generation reactors as, say, binding energy extractors (BEEs) on the model of what medical equipment manufacturers did about 1979. Magnetic resonance imaging (MRI) avoided the long-standing scientific name, nuclear magnetic resonance (NMR), probably because the marketing people warned that including the word "nuclear" was a downer.

	Ability to expand	Public view	Down side	Ups & Downs	Foot print	Storage needed	Enough by 2020?
Third Gen Nuclear	10X	caution	many	steady	mining	spent fuel	●●● to ●

E ven if the developed countries bring their addiction under control, fossil fuel use has soared in the rest of the world. Unless we can provide an alternative to burning coal and oil, they won't change their ways fast enough. If we can convert them to using electric or compressed-air vehicles, then the issue becomes clean and cheap electrical power. In the long run, in-country deep geothermal might be best. For 2020, we need an additional, sure-fire strategy.

Fifteen wires to move 10 gigawatts with high-voltage AC, lose 10% by 1,300 km

Four wires to move 10 gigawatts with high-voltage DC, lose 10% by 2,100 km

One way of solving this would be connecting all countries to regional power grids constructed with efficient DC transmission lines. It's an old technology commonly used for underwater and underground power lines; the aerial versions of DC are now used for power lines over 1,300 km long. (That's the length of the DC line from the Washington-Oregon border down to Los Angeles.)

The longest DC transmission line in the world, completed in 1983, spans 1,700 km in the Congo. A 3,000 km DC line from Spain would cover all of northern Africa; one from Johannesburg would cover all of southern and eastern Africa plus Madagascar; one from Mexico could cover the Caribbean and into South America; one from Hong Kong or Australia could cover southeast Asia.

This would enable nuclear power plants to be restricted to the present thirty-one countries. That's important if, rather than waiting for the fourth-generation BEEs, we are to use the current generation reactor designs that incidentally yield bomb material.

The architects' sensible plans for green buildings are long-term only, unable to help much in closing The Gap by

2020. It's the same for rapid transit. I'm inclined to encourage their growth but put the big money elsewhere for now. Our enthusiasm for long-term thinking is, sad to say, short sighted given the 2020 emergency. What we do for 2020 will reframe the problem, and new science and technology by then will hopefully show us a better path.

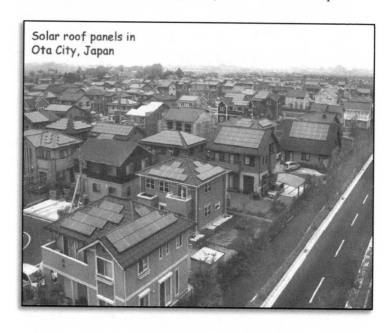

Solar roof panels in
Ota City, Japan

The growth in solar panels since 1995 has been impress-ive. Solar currently provides about 1 percent of world electricity (much less in the U.S.). The photovoltaic version is especially suitable for off-grid use in modernizing countries.

"Concentrating solar" heats a fluid that runs the usual steam turbine. It's being tried out in sunny Spain with mirrors that track the sun, keeping sunlight focused on the top of the tower. Both solar versions have unpredictable ups and downs as the clouds move by. Solar is also used for direct (no electrical middleman) heating, such as rooftop hot water heating.

Concentrating solar array near Seville, Spain

Hydropower is the current big item (after nuclear) in the C-free power portfolio and efficiency improvements can be made by modernizing existing dams. The number of dams in the world grew from 5,000 in 1950 to more than 45,000 today—that's two dams a day for 50 years—but it is close to saturated. Low-rise and stream-flow hydro are not going to be big players for 2020.

Biofuels, however green in small amounts, turn out to be a bad idea when scaled up. First, a serious drought (and in the coming decades, they are very likely) would impact both food supply and transportation fuel simultaneously. All prices would soar and the economy would stagger. As any investment advisor will tell you, spread your bets to avoid simultaneous downturns. Hydro power is already at risk in a drought and we should be adding drought-resistant alternative power, not biofuels.

Biofuels in developing countries will also require more land clearing, reducing the world's carbon sinks and depleting poor tropical soils—as is already happening with "deforestation diesel." European subsidies prompted an enormous boom in planting palm oil trees in Indonesia and Malaysia. Cutting the forests and draining the swamps emitted far more carbon than could ever be saved from using biodiesel.

My take:

	Ability to expand	Public view	Down side	Ups & Downs	Foot print	Storage needed	Enough by 2020?
Solar	lots	OK	**few**	night, clouds	multi use	some	●●
Wind	lots	ugly	noise, bird kills	fickle, unstable grid	multi use	some	●●
Biofuels	compete with food	organic fuel	not C-neutral	drought	huge	some	●
Flow & tidal hydro	some	caution	ecology	drought	large		unlikely
High-rise Hydro	nearly full	nice lakes	dam failure	drought	large	lakes	none

My father ran a medium-sized insurance company in Kansas City in the 1950s, back when fire departments fought a lot more home and building fires than they currently do. When we were driving around town in my youth, he was always pointing out everyday situations that had gotten some people into big trouble. (And so I grew up naïvely supposing that it was standard practice to routinely estimate risks and take sensible precautions!)

Indeed the reason that there aren't as many fires these days is because society has incorporated into building codes and regular inspections what the fire chiefs and insurance executives had noticed over the years.

Later, twenty years of talking shop with the neuro-surgeons every day helped to form my notions about when you can afford to wait and when prompt inter-vention is needed. James Lovelock, Jared Diamond, and I are all Ph.D. physiologists who, during decades of medical research en route to looking at things more broadly, also learned to think like physicians.

Note that both my father and my neurosurgical colleag-ues were at the top of a pyramid of information. For example, few people in medicine forty years ago really suspected how dangerous it was to ride a motorcycle with-out a helmet. But the neurosurgeons were the ones who had to cope with the broken heads and they realized the protection that the helmet conferred. This gave them the responsibility to do something, to try and prevent the ruined lives. So they pushed for better helmet designs and for laws that required helmets to be worn. Earlier they and other physicians had done the same thing for seat belts.

It used to be that you had to be a scientist in order to realize how serious the climate problem was becoming. You needed the view from the top of that pyramid of information. Now anyone who can read a book on global warming or watch a documentary film can gain much of that formerly rarified view.

Consider for a moment your present situation. You are now better informed about climate than thousands of your neighbors. What can you do with that knowledge?

For myself, I recall the moment which led to this book—a sinking feeling when it finally became clear that there was a 2020 emergency developing. It felt like what many have described for the eve of a great war, where future plans have to be put on hold, superseded by civic duty. It becomes payback time. I realized, as Tim Flannery put it, that "in the years to come, this issue will dwarf all the others combined. It will become the *only* issue."

Even the well-informed politicians, who understand the actions needed, will require reassurance that starting a major makeover won't result in budget-conscious voters throwing them out of office at the next opportunity. (In the U.S., there is a perception that this happened in 1980 and 1994, following modest energy initiatives.) So serious political action on an energy re-make may need an over-whelming advance endorsement, repeated over and over—not just an initial expression of concern on your part.

My advice would be to set a good long-term example for the kids and developing nations, but don't count on it solving our big 2020 problem in time. Remember that the real focus needs to be on political action to stop this runaway train, real soon.

Ranking the Major C-free Candidates
for stopping emissions growth by 2020

		Ability to expand	Public view	Down side	Ups & downs	Foot print	Storage needed	Enough by 2020?
E **N** **E** **R** **G** **Y**	**Hot Rock Energy**	huge	Just another well?	Month of small EQs?	**very stable**	**very small**	none	●●●● to ●●
	Nuclear	10X	caution	many	steady	mining	spent fuel	●●● to ●
	Solar	lots	OK	**few**	night, clouds	multi use	some	●●
	Wind	lots	ugly	noise, bird kills	fickle & unstable grid	multi use	some	●●
	Biofuels	compete with food	organic fuel	not C-neutral	drought	huge	some	●
	High-rise Hydro	nearly full	nice lakes	dam failure	drought	large	lakes	no
	Coal but capture the CO2	large	caution	storage burp	steady	67% more coal	huge	no
O **T**	Plankton iron blooms	large	caution	side effects?	likely	fleet of ships	some	●
H **E**	Plug-in hybrid cars	large	**very good**	mining battery				●●●
R **S**	Compressed air car	large	none yet	air tanks				●
	Improve efficiency	good	in favor	slow grind	—			●
	Energy Diet	limited	a pain	easy to fail	yo-yo			no

Not comprehensive. Opinionated. Likely outdated (updates at *Global-Fever.org*).

Fossil fuels helped us to fight wars of a horror never contemplated before, but they also reduced the need for war. For the first time in human history - indeed for the first time in biological history - there was a surplus of available energy. We could survive without having to fight someone for the resources we needed. Our freedoms, our comforts, our prosperity are all the products of fossil carbon, whose combustion creates the gas carbon dioxide, which is primarily responsible for global warming.

Ours are the most fortunate generations that have ever lived. Ours might also be the most fortunate generations that ever will. We inhabit the brief historical interlude between ecological constraint and ecological catastrophe.

—commentator George Monbiot, 2006

Before, if we screwed up, we could move on. But now we don't have an exit option. We don't have another planet.

—climate scientist Will Steffen, 2005

Economists may tell you that it will take twenty years—but when there is a war on, you get it done in a few years.

—biologist Tim Flannery, 2007

20

Arming for a Great War

In *Collapse*, his survey of past societies that failed and succeeded in dealing with collapse-threatening environmental change, Jared Diamond identified three stages in avoiding collapse.

First, it took each society a while to identify that there was a problem. For our present carbon creep, this realization happened between 1938 and 1965 for the relevant scientists, long after the creep began around 1750. In terms of identifying that we are already in dangerous territory for experiencing abrupt climate shifts, I'd guess it dawned on many climate scientists between 1998 and 2005.

Second, Diamond observed, comes the discovery of some ways of coping with the problem. Cap-and-trade, the cornerstone of the 1997 Kyoto agreement, ironically arose out of the success of the U.S. experience with tackling smokestack sulfur ("acid rain") emissions using cap-and-trade.

Third is actually doing something about the problem. Europeans made effective starts before 1980. But it will now take about six times the Kyoto goals just to cancel out the subsequent increases in yearly emissions from China and the United States. And Kyoto was an inadequate goal.

Now we need that first turnaround accomplished before 2020.

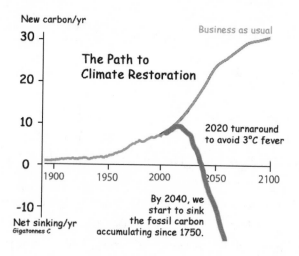

Can we do so much by 2020? You only have to look back to see great spurts of scientific and technological progress. Who in 1992, trying out the first web browser via a phone line modem as I did, would have thought that the web would expand so fast? It went from one page to a billion in only four years, indexed by free search engines.

The first industrial revolution, spanning the 1770s to the 1860s, saw the earlier scientific inventions such as the battery extended to create the telegraph and electric motor. The steam engine led to steamboats and locomotives. Photography. Anesthesia. Refrigeration.

Then progress sped up. The second industrial revolution from the 1870s to the 1910s built on this industrial base and the mid-nineteenth-century discoveries in physics and chemistry. And so in its first twenty years we got the telephone, the internal combustion engine, the light bulb, hydroelectric dams, the linotype machine, and a primitive car. Science surged again, producing the germ theory of disease, genetics, and artificial vaccines.

In the next twenty years, 1890 to 1910, we got the theory of relativity, the photoelectric effect, the first textbook of psychology, and such marvels as neurosurgery, motion pictures, air conditioning, airplanes, and radio.

In the 1940s and 1950s, we discovered the jet stream, DNA, and global warming. We invented the computer, nuclear power stations, the transistor, fiber optics, satellites in orbit, the Pill, antibiotics, and major psychoactive drugs. Polio vaccines saved millions the prospect of life in an iron lung. Television surged. Films were in color.

In the 1960s and 1970s, we discovered continental drift, put communication satellites in geosynchronous orbits, went to the moon, did heart transplants, invented the Internet, and created personal computers. Scientists started using email and spreadsheets.

A lot can happen in only twenty years—especially with our current scientific momentum. We cannot rely on future progress to save the situation, but there are good reasons to think that some parts of the problem may prove tractable, given enough effort.

Next, consider the surprises that might interrupt our orderly efforts to turn around and restore the climate.

Suppose storm surges from ocean warming scour those vulnerable coastal cities that were slow to build surge barriers. Besides all of the loss of life, there will be an enormous expense from rescue, temporary housing, and rebuilding. Then there will be a crash effort to construct sea walls and surge barriers. It will be the New Orleans problem, repeating in other coastal cities. Will there be time and C-free energy in sufficient quantities to do both that and clean up the CO_2?

Or suppose that floods and droughts create a large homeless population that moves across national borders? Or that an unpaid hungry army goes foraging in the country next door—and the reaction starts a war? There will surely be a lot of these unhappy events. Should charismatic leadership arise, something like what we are now seeing in the radical jihadists, terrorists might effectively disrupt the developed countries to exact "relief aid" in a classic protection racket. And so the immediate demands of war would again override the necessary environmental infrastructure investments.

A similar effect would occur from revolutions. A major country might descend into civil war for any of the usual reasons. New reasons for civil war might include frustration with a government that is too slow to act on climate ("Our laws are not a suicide pact!") or intergenerational conflict between those guarding their assets for an uncertain retirement and the asset-poor younger generation, insisting on taxes for expensive infrastructure improvements.

Whatever the cause of a schism, countries take many decades to recover from such national traumas and that will be time lost—time that the world cannot afford when trying to head off runaway climate change.

Before 2040, we need to reduce net CO_2 additions to zero—and begin reversing the CO_2 concentration back toward 1939 values. (I pick that year only because that's when I began personally adding to the CO_2 problem.) This will only happen if the technology of the developed world has become good enough to compensate for what's still going on in failed nations and others too disorganized to get their energy act together.

Wisdom, it is said, requires an ability to understand human nature, perceive a situation clearly, and decide despite ambiguity and uncertainty. It requires an ability to get past the common roadblocks, such as denial. By this standard, much of the political leadership is not wise when it comes to climate change.

Remember that foresight is not about predicting the future. Rather, foresight allows you to mold the future. We'd better hurry implementing that foresight because positive feedbacks seem to have put the climate machine into Fast Forward.

We are already in dangerous territory and have to act quickly to avoid triggering widespread catastrophes. The only good analogy is arming for a great war, doing what must be done regardless of cost and convenience.

If flying-saucer creatures or angels or whatever were to come here in a hundred years, say, and find us gone like the dinosaurs, what might be a good message for humanity to leave for them...?
WE PROBABLY COULD HAVE SAVED OURSELVES, BUT WERE TOO DAMNED LAZY TO TRY VERY HARD . . .

—author Kurt Vonnegut

As the effects of global warming become more and more apparent, will we react by finally fashioning a global response? Or will we retreat into ever narrower and more destructive forms of self-interest? It may seem impossible to imagine that a technologically advanced society could choose, in essence, to destroy itself, but that is what we are now in the process of doing.

—science writer Elizabeth Kolbert

The only certainty is that we have to act. How could I look my grandchildren in the eye and say I knew about this and I did nothing?

—naturalist David Attenborough

The biggest challenge is how to get people to wake up and realize this is a one-shot deal. If we fail, we are witting participants in the biggest experiment that humans have ever done: moving CO_2 levels to more than twice their value in the past 670,000 years and hoping it turns out okay for generations to come.

—chemistry professor Nathan S. Lewis

21

Get It Right on the First Try

As Samuel Johnson said in 1777, "When a man knows he is to be hanged in a fortnight, it concentrates his mind wonderfully." So let me reframe our turnaround-by-2020 emergency by rearranging some of the prior points.

Our global fever has been spiking for the last thirty years, thanks to an accumulating blanket of CO_2 around the earth. Treatment proposals range from a low-carbon diet to elaborate umbrellas in space. Most will fail to do the job in time: too little, too late, or too local.

That's because there is a big difference between a fever that peaks at about $2^{\circ C}$ [$3.6^{\circ F}$] above the 1990 global average and one that peaks at $3^{\circ C}$ [$5.4^{\circ F}$]. That additional fever has been recently recognized by climate researchers as the difference between an outcome that is bad and one that is terrible.

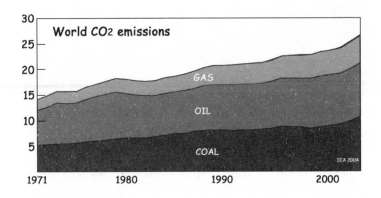

Holding it below $2^{\circ C}$ requires, before 2020, stopping the annual escalation of emissions, recently at 3 percent each year (it was 1.1 percent in the 1990s but developing countries such as China and India have begun to burn a lot of coal and oil). If this annual increase in the burn rate doesn't go to zero and turn around until 2040, we get the $3^{\circ C}$ fever.

That isn't *agree* to do something by 2020, it's *accomplish* it before 2020. This time frame rules out waiting for some equitable international agreement to be negotiated or relying on slow half-measures such as cap-and-trade to encourage the free market to reorganize the energy sector.

Unlike the downwind fallout of coal ash and acid rain, our CO_2 mixes with everyone else's within a year, then hangs around for centuries like a shroud. The less organized countries, by modernizing like we did, can easily cancel out our greenhouse efforts. And so we must provide them with inexpensive solutions, ones that keep them from burning their local coal (the worst and cheapest of the fossil fuels) for electricity or cutting down their remaining forests to grow biofuels.

The time for gradual solutions and new inventions is largely past, thanks to our party-now-pay-later attitudes since scientists began warning us of the growing problem in 1956. As a consequence of our Faustian bargain, that $2^{\circ C}$ fever will kill off not just the coral reefs and polar bears but (in my opinion) half of all plant and animal species. That's how bad our stewardship of the Earth has become.

Fortunately, we already have the technology in hand to head off the far more terrible consequences of a $3^{\circ C}$ fever, a world full of refugees and genocides, Darfur writ large. No single treatment is both quick and global but there is one combination of existing technologies that might head things off at difficult before they advance to terrible. They're my suggestion for a "Turning Around by 2020" agenda.

First, we need to convert most vehicles to C-free energy. Plug-in hybrids will shift the transportation sector's energy needs from oil to whatever produces the local electricity. This need not mean batteries on board as the electrical power can be used to create some other intermediary fuel, hydrogen for fuel cells or compressed air for an air car.

This will get rid of much of the CO_2 from gasoline. Even if the electricity comes from coal, there's a gain because of size efficiencies and avoiding the waste of idling gasoline engines in traffic jams.

There is much talk of expanding solar, wind and biofuel energy sources for the long term. But we cannot count on such developments succeeding in the short run of 2020. While five times as many nuclear plants could replace most coal use in the U.S., solar and wind would need to increase more than a thousand times to do the job. For geothermal, the increase needed would be a hundred times—but it might be quicker than nuclear.

I cannot believe that the other renewables, or retrofitting thousands of coal plants to capture and store the CO_2, are capable of being ramped up fast enough for the 2020 time frame. The only well-developed clean power sources are geothermal, hydro, and nuclear—and high-rise hydro is pretty well maxed out.

Second, for the short run we need to build a great many Hot Rock geothermal power plants or clone a smaller number of nuclear plants. We're just getting started with Hot Rock Energy production and, even though all the parts and pieces are familiar, we cannot rely on the new combination yet in quite the same manner as we can for expanding nuclear. So let me discuss the traditional objections to nuclear in the light of twentieth-century experience.

Despite the levels of public concern, nuclear power has the best safety record, by a large margin, of any means of generating electricity—even hydro, as dams fail regularly somewhere in the world. Compare less than one fatality per year for nuclear to the monthly headlines of yet another mining disaster. In China alone, 6,000 coal miners die each year. Longer-term risks from nuclear are also tiny compared to the cancer and asthma risk of living down-wind of coal-fired plants—as does everyone in the eastern half of the U.S.

It takes about five years to actually construct a new nuclear plant, once permits are lined up. No new nuclear plants have been started in the U.S. since 1978 (nor have we reprocessed any of their spent fuel as France, Russia, and Japan do).

France has switched to nuclear for 78 percent of its electricity. Hydro gives it another 13 percent. So France is 91 percent clean, 9 percent dirty—and Texas is the exact opposite. Texans now get 91 percent of their electricity from fossil fuels, almost twice the national average.

If France switched its vehicles over to electrical power, it would serve as an even better C-free energy model for the world. Much as I admire Denmark's style of distributed cogeneration and the move to renewable wind and solar energy, there simply isn't time to scale that up around the world before 2020, given how many coal trains and oil tankers need to be retired quickly.

Too much of the present discussion is either pie-in-the-sky or envisages a slow evolution of urban architecture, commuting, energy, and civic virtue. We once had time for such planning. We can still explore on many fronts at once but for 2020 we need to be quickly building something foolproof. For the heavy lifting, that looks like Hot Rock geothermal or third-generation nuclear.

For speed, we now have to go with what we've got. We can still ramp up the others for the long run. Each time they add up to 1,000 megawatts, we can cancel an order for a nuclear plant.

Does substituting electricity for gasoline and ramping up nuclear and geothermal solve the world's 2020 problem? Alas, you can't simply give a nuclear power plant to a developing country where plant security and the rule of law are fragile things.

So the third thing we must do is quickly build networks of the low-loss DC transmission lines, where it takes 2,100 km to get a 10 percent line loss (a mere nothing compared to the 67 percent energy loss for generating the electricity from fossil fuels). They will allow continent-wide distances between power plant and consumer. Thus, DC will allow the developed countries on each continent to house the nuclear reactors and provide for both the electricity and transportation needs of the developing countries. The price would need to be subsidized to keep them from burning their coal or importing oil. If Hot Rock geothermal suits, we can start drilling deep wells for them.

Note that most of the candidate technologies are more than fifty years old. You can't say that we don't have the technology to solve the 2020 problem.

That's my three-part suggestion for how to achieve "Turning around by 2020." Clearly we'll also need C-fees to encourage replacing vehicles, forming car pools, and remodeling buildings. We'll need some cap-and-trade to encourage innovation more broadly, though it has all the potential problems and delays of inventing a new international currency and auditing credits for performance. We dare not rely too much on a low-carbon diet or individual initiative. That's because diets often fail within several years of an initial "success."

To go further and actually lower the greenhouse fever, we will need to take much more CO_2 out of the air than we put in. That's necessary because CO_2 serves to acidify the surface layer of the ocean—which in turn will reduce photosynthesis, crash the food chain, and kill the carbon pump which makes limestone out of CO_2.

"Inevitable" is a much overused word. The only thing inevitable is that nearly all extinct species will remain gone forever. Success in lowering global fever won't reverse all of climate change but extreme weather and desertification ought to turn around with the temperature drop. And while we speak of the first degree of fever as the inevitable delayed consequences of past emissions, much of this could also be headed off if we were able to remove CO_2 from the air with a large scale enhancement of C sinks.

Even though China now emits more CO_2 annually, the U.S. has contributed more, over the years, to the insulating blanket around the Earth than any other country (30 percent). Yet the U.S. still resists even baby steps toward energy reform. The rest of the world sees the U.S. as the 500-pound gorilla. (The other gorillas won't go very far down a path until the big guy finally gets up and comes along.) A colleague of mine came back from an international meeting of climate scientists having overheard someone say that, since he didn't live in the U.S., he felt impotent to deflect the world from the road to catastrophe.

I've emphasized a quick technofix because the window of opportunity is closing on us and we've only got one planet to lose. Time is short. And because of starting late, we must get this right on the first try or we may be trapped in a runaway.

That's where we lose existing sinks for CO_2 via closing the leaf pores, drought, fire, baking the soil, and ocean acidification. Some carbon sink regions turn into net sources, as when wood decay emits more CO_2 than the remaining trees can recapture. We already see that in the tropical rain forests during an El Niño. That pushes up the fever and further accelerates the decay in topsoils, which add more CO_2. This accelerates the temperature spike. Like the squeal of a public address system when the speakers are heard by the microphone, it just keeps building up, bringing most activities to a shrieking halt.

That's why planting more forests is not a very reliable way to sink CO_2.

I will leave the downside scenario to the producers of disaster movies. But should we find ourselves in the midst of a runaway climate change, there will still be much meaningful work to do, such as the organized kindness of relief work. There will also be things to do for a future civilization.

Our civilization presumes that each generation has a responsibility to make the future better for the next. There is a possibility that our civilization will fail. Catastrophically. And so we would want to make sure that the downsized remnant of our society has a good Recovery Manual to guide the eventual restoration of civilization.

There's no point in them having to make the same mistakes as we did. Historians will outline how to make a transition from autocratic rule to democracy, how to keep peace between minority groups, how to keep a free flow of information and opinion, and how to recognize disinformation and fear campaigns.

The health care professionals will need to write do-it-yourself medicine books for a world with minimal medical supplies, as well as crafting recipes for bottom-up production of the major medicines and dental tools. Technologists will need to describe how to bootstrap energy production and manufacturing.

Since we cannot assume working technology to read CDs and the like, the operations manuals will need to be in durable book form. Cartoons on monuments will direct finders to mountain tops where they will find caches of books on decay-resistant media and directions to large libraries buried in dry caverns.

But we'll need insurance against book burners. Since salvaging scrap metal will be a major endeavor for survivors, perhaps we will photoengrave small-print chapters of the Recovery Manual on every commonplace metal surface—airplanes, cars and trucks, the inside of beverage cans, and the skins of light rail cars. Someone will start reading the cans. Piecing together the jigsaw puzzle to get a full document will become a community effort.

If the recovery books turn out not to be needed because the technofix succeeds, the books will nonetheless prove useful for self-empowerment in developing countries. And, in my experience, the effort to boil down the material to fit a smaller space often results in making new discoveries about how things hang together. We might discover a few things from making this effort.

There is something to be said for starting such projects immediately. They should be done cheerfully, as in the claim that carrying an umbrella keeps it from raining. The recovery manuals would provide meaningful work for those who cannot participate more directly in climate restoration. Most importantly, the effort might bring home the seriousness of our situation to voters confused by the real choices.

We need to remember the lessons of behavioral economics as, alas, they all work against taking action on climate. It's possible to work around them, but only if you know they are there.

People "endow" their possessions and paychecks with inordinately high value, simply because they possess them. People feel the pain of a loss more acutely than the joy of a gain—one reason why future gains are hard to balance with the loss of present-day spending money.

Status quo bias, the tendency to keep doing what you've always done, is often stronger than the rational arguments for changing course.

When dealing with an unfamiliar situation where it is difficult to discern the true costs, there is a tendency to seize upon any big number as a meaningful starting point. This provides an "anchor" for the discussion. To frame the discussion, be careful which anchor you choose.

People have a tendency, when faced with too many choices, to decide *not to decide*. Disinformation campaigns about climate science have been designed to promote decision paralysis and I hate to think of what new disinformation campaigns will be deployed by the highly profitable coal and oil interests when they contemplate their sales dropping by half.

Hopefully they will realize that they can maintain profits by doubling the price instead of sowing confusion. Still, we'll have to prevent them from again lowering prices for long enough to bankrupt the alternative fuels entrepreneurs. We can do that by establishing a base price where taxes kick in to make up for a lower price.

Now in conclusion, a few words addressed to the comfortably middle aged.

This isn't a problem that you can leave for the next generation to solve.

Your age and experience puts you, for better or worse, in the driver's seat for another decade or two. Unless you slow down this runaway coal train, it is going to smash us.

And, if you carry civilization through this difficult time, your generation will be known as the can-do generation of all time.

Finally, some perspective for the up-and-coming generation of students and young professionals. Remember that the leadership of the civil rights movement was young. They accomplished a great deal within fifteen years. Martin Luther King, Jr., was, at age 35, the youngest person to ever win the Nobel Peace Prize. Trashing the planet is a great moral issue and it needs great leadership.

Your generation gets to do the real makeover of our civilization. Many people will soon concede that real changes must be made, that we cannot stay trapped in business as usual.

Avoid being distracted by the doom-and-gloom types who see each hurricane as a sign that the apocalypse has come. You will also be competing with the gut-feeling, do-nothing happy talk that slides right past the problems—and the opportunities to intervene.

Others will hinder you more directly. There will be those who will want to increase the burn rate for resources on the grounds that The End is Near (James Watt, the U.S. Secretary of the Interior under President Ronald Reagan, had exactly such an attitude).

Even worse are the fans of Armageddon, some of whom will be a real danger (Aum Shinrikyo, in their 1995 sarin attacks in the Tokyo subways, was trying to promote a world war that would destroy everyone—except, of course, the faithful).

But no previous generation has ever had such an opportunity as you now have. As you tackle the longer-term issues of redesigning our civilization for resilience, you may be able to reconfigure education, governance, and the social contract. You will be laying the reinforced foundation for the next thousand years of our civilization.

Our present civilization is like a magnificent cathedral, back before flying buttresses were retrofitted to stabilize the walls. Civilization now needs such a retrofit. It will be a large undertaking, not unlike those that once went into building pyramids and cathedrals. I'm optimistic that the younger generation can create a better civilization during our major makeover—provided that those currently in the leadership can stop this runaway coal train, real fast.

Climate change is a challenge to the scientists but I suspect that the political leadership has the harder task, given how difficult it is to make people aware of what must be done and get them moving in time. It's going to be like herding stray cats, and the political leaders who can do it will be remembered as the same kind of geniuses who pulled off the American Revolution.

The public interest requires doing today those things that men of intelligence and goodwill would wish, five or ten years hence, had been done.

—Edmund Burke (Anglo-Irish statesman, 1729-1797)

We are now faced with the fact that tomorrow is today. We are confronted with the fierce urgency of *now*. In this unfolding conundrum of life and history there is such a thing as being too late. Procrastination is still the thief of time. Life often leaves us standing bare, naked and dejected with a lost opportunity.

The 'tide in the affairs of men' does not remain at the flood; it ebbs. We may cry out desperately for time to pause in her passage, but time is deaf to every plea and rushes on. Over the bleached bones and jumbled residue of numerous civilizations are written the pathetic words: 'Too late….'

—Martin Luther King, Jr., 1967

Eyes from the past. The big statues of Easter Island originally looked like this one. The eyes were often looted by raiding tribes. Archaeologists found this pair hidden nearby and restored the toppled top hat as well.

Most societies that collapsed in the past did not realize that they were slowly destroying their environment. As Jared Diamond describes in his book *Collapse*, a few pulled themselves back from the brink — but not the one on Easter Island.

Acknowledgements

My thanks to the many climate scientists who have patiently answered my questions over the years, starting with Stephen Porter and Hans Oeschger in 1984. Among those who really opened up new avenues of thought for me are Richard Alley, David Archer, Richard Gammon, Ed Miles, Ray Pierrehumbert, Stefan Rahmstorf, Peter Rhines, Stephen Schneider, Eric Steig, and Tom Stocker.

My beta testers included Katherine Graubard, Susan Rifkin, Kathryn Moen Braeman, Linda Mitchell, Peter Rockas, Carol Nygren, and Susan M. Johnston. The readers have them to thank for flagging the bumps in the road and alerting me to metaphors that didn't work. For pointing out authorial excess, there is nothing quite like having both a wife and an ex-wife who are accustomed to wielding an editor's red pencil.

For helpful comments, I thank Deborah Gardner, Mike MacCracken, Barrie Pittock, Sarah Das, and Sir Crispin Tickell.

A book like this is possible only because the earth sciences community has been at the forefront in providing web access to their research discussions and teaching materials. I first remarked on this back in 1997 when writing my *Atlantic Monthly* article. It is even truer today, what with sites such as *RealClimate.org* where serious questions get serious answers. I thank them all.

About—Polar Ice

New York Times
24 October 1954

How fast does it melt? An expedition seeks the answer.

By BENJAMIN POWELL

ONE of the objectives in the forthcoming American expedition to Antarctica is to check on the status of the stupendous ice fields and glaciers of that little-known continent. The particular point of inquiry concerns whether the ice is melting at such a rate as to imperil low-lying coastal areas through raising the level of the sea in the near future. There is no saying as yet how imminent that threat might be, but there is little question that there is enough of the frozen stuff around to cause trouble if it all melted suddenly.

IN THE SOUTH

The Antarctic Continent is about 5,000,000 squ[...] area, and [...] onl[...] fr[...]

[...] of [...] not be likely [...] great effect if it [...]

HOW MUCH DANGER?

A greater potential danger (nevertheless remote) is the Greenland Ice Cap. Here is an enormous, rounded hump of ice, piled over the mountains and valleys of the world's largest island. The ice covers about 700,000 square miles, leaving a sparse fringe of exposed rock around the coast. The thickness is believed to average about 1,000 feet, but soundings have indicated depths of 10,000 feet or more.

Elsewhere in the northern hemisphere are great deposits of ice and snow in the Himalayas, Alaska, the Alps and the Scandinavian mountains.

There is no doubt that the ice of the far north is melting faster than it is replaced by new snowfalls (actually most of the Arctic gets only about 8 inches of precipitation a year). Glaciers all around the northern world are receding at a clearly measurable pace, and have been for several generations. Yet the rate of melt is so slow as to leave no rise in tide levels sufficient to alarm humans. Greater solar radiation is given as the cause of the recession, but greater solar radiation also increases the rate of evaporation.

Assuming a very rapid, almost instantaneous melting of Greenland's ice alone, it has been estimated that the level of the oceans would be increased about twenty-five feet. But, scientists point out, conditions bringing that about would be of such a cataclysmic nature that there would probably be no one left around to do the worrying, anyway.

An ice breaker off Ross Barrier.

Read Widely

This section and the chapter notes are also available at *Global-Fever.org*, augmented with live links.

When undertaking this book on our global fever, I decided to write a cheerful book in parallel. It became *Almost Us: Portraits of the Apes*. You might wish to employ a similar back-and-forth strategy when reading more about our big problem.

If you haven't already, I'd suggest reading

Jared Diamond. *Collapse: How Societies Choose to Fail or Succeed*. Viking, 2005. Anthropology and ecology of past societies, not specifically global warming but the essential stage-setting. Excellent, as usual.

Al Gore. *An Inconvenient Truth*. Rodale Press, 2006. Closely follows the world-famous film that won an Oscar. As Jim Hansen says, "Al Gore may have done for global warming what 'Silent Spring' did for pesticides."

Then consider reading one or more of these books:

Robert Henson. *The Rough Guide to Climate Change*. Rough Guides, 2006. Don't be fooled by the travel-books connection. It's one of the best of the reader-friendly books that could also be used for climate courses. The author is a science writer at the National Center for Atmospheric Research in Boulder, Colorado.

Mark Lynas. *Six Degrees*. Fourth Estate, 2007. His chapter on the consequences of a 1°C fever is sobering enough, but then he works his way through the consequences of the 2, 3, 4, 5, and 6° fevers and "Choosing our future." Very well done. It also shows that with a first-class honours degree in history and politics, you can read and understand much of climate science.

Joseph J. Romm. *Hell and High Water*. William Morrow, 2007. An excellent book of climate science plus advocacy by a former acting Assistant Secretary in the U.S. Department of Energy. The author is a Ph.D. physicist and oceanographer by training but his father was a journalist—and it shows.

Then consider these when branching out:

Brian Fagan. *The Long Summer: How Climate Changed Civilization*. Basic Books, 2004.

Tim Flannery. *The Weather Makers*. Atlantic Monthly Press, 2005.

Ross Gelbspan. *Boiling Point*. Basic Books, 2004.

Elizabeth Kolbert. *Field Notes from a Catastrophe*. Bloomsbury, 2006.

Eugene Linden. *The Winds of Change*. Simon & Schuster, 2006.

James Lovelock. *The Revenge of Gaia*. Penguin/Allen Lane, UK, 2006.

George Monbiot. *Heat: How to Stop the Planet Burning*. Penguin/Allen Lane, UK, 2006.

Fred Pearce. *The Last Generation: How Nature Will Take her Revenge for Climate Change*. Eden Project Books, UK, 2006.

A. Barrie Pittock. *Climate Change: Turning Up the Heat*. CSIRO, Australia, 2005.

Phillip W. Schewe. *The Grid*. Joseph Henry Press, Washington DC, 2007.

Spencer R. Weart. *The Discovery of Global Warming*. Harvard
    University Press, 2003. Updated version at
    www.aip.org/history/climate.

On the web, I would initially avoid search engines because of
the disinformation problem for climate matters. Try

RealClimate.org, done by real climate scientists,

Society for Environmental Journalism at
    www.sej.org/resource/index18.htm.

Professor Stephen Schneider's climate website,
    stephenschneider.stanford.edu,

American Institute of Physics,
    www.aip.org/history/climate/links.htm

Pew Center on Climate Change, *www.PewClimate.org*,

Climate Institute at *Climate.org*,

ClimatePrediction.net

The National Center for Atmospheric Research at
    www.ucar.edu/research/climate/future.jsp.

UK's Hadley Centre at
    www.metoffice.gov.uk/research/hadleycentre/.

Union of Concerned Scientists at *ClimateChoices.org*.

Rocky Mountain Institute at *www.RMI.org*.

World Resources Institute, at *WRI.org*. Their *Navigating
    Numbers*, by Kevin A. Baumert, Timothy Herzog, and
    Jonathan Pershing, is quite useful.

BBC's updated climate pages at *www.bbc.co.uk/sn/hottopics-
    /climatechange/*

New York Times at *topics.nytimes.com/top/news/science/topics-/globalwarming/index.html?8qa.*

American Association for the Advancement of Science at *www.aaas.org/climate/*

They all have a list of recommended links to other sites, regularly updated. The Society of Environmental Journalists has an excellent list of lists for all sides of climate change at *www.sej.org/resource-/index18.htm*—it even includes the Birdwatcher's Guide to Global Warming!

Armed with some of the science, you can gradually branch out to the wider web. See how quickly you can spot the front organizations for the not-a-problem promoters of business as usual through more delay. Most of them have invented fancy names for themselves in order to slip past your guard; most include some good science to help disguise their propaganda. Once you are good at it, test your skills at *GlobalWarming.org.* See the Union of Concerned Scientists' 2007 report on ExxonMobil's $23-million attempt to mislead the public at *ucsusa.org/assets/documents/global_warming/exxon_report.pdf.*

More advanced readers should take a look at

Intergovernmental Panel on Climate Change, *2007 Summary for Policymakers* for each of the three working IPCC groups, at *www.ipcc.ch*. There are also Technical Chapters with all the references up to late 2005.

U.S. Global Change Research Program, *Climate Change Impacts on the United States,* USGCRP, 2001 *www.usgcrp.gov/usgcrp/Library/nationalassessment/*

The Arctic Council, *Arctic Climate Impact Assessment,* Arctic Council, 2005. *www.acia.uaf.edu*

James Hansen, "A slippery slope," *Climatic Change* 68 (February 2005): 269-279. At *dx.doi.org/10.1007/s10584-005-4135-0.*

Illustration List

Most of these illustrations may be freely borrowed for non-commercial and educational uses. They may be downloaded from *Global-Fever.org*.

Notes

The following notes are also on the web at *Global-Fever.org*, referenced by this book's page numbers and including web links to many of the citations. Author (year) citations usually refer to a book in the **What to Read** section.

vii Munch lithograph at *www.jesuits.ca/orientations/Munch2.html*. Letter from *en.wikipedia.org/wiki/The_Scream*.

<table>
<tr><td>Chapter 1.</td><td style="text-align:right">The Big Picture</td></tr>
</table>

3 Guy Callendar, "The Artificial Production of Carbon Dioxide and Its Influence on Temperature." *Quart. J. Roy. Meteorol. Soc.* 64 (1938): 223–40.

6 Coal twice as bad, per kilowatt-hour generated: *en.wikipedia.org/wiki/Energy_content_of_biofuel*.

7 These estimates do not include the effects of tropical forest fires on carbon emissions, which are much more difficult to measure. When the 1997/98 El Niño episode provoked severe droughts in the Amazon and Indonesia, large areas of tropical forest burned, releasing 0.2 to 0.4 Pg of carbon to the atmosphere. If droughts become more severe in the future through more frequent and severe El Niño episodes, or the dry season becomes lengthier due to deforestation induced rainfall inhibition, or there are rainfall reductions due to global warming, then substantial portions of the 200 Pg of carbon stored globally in tropical forest trees could be transferred to the atmosphere in the coming decades. Global carbon emissions from fires during 1997/98 El Niño are estimated at 2.1 ± 0.8 PgC and South and Central America contributed ~30% of global emissions from fires. See *www.joanneum.at/Carboinvent/post2012_/Bird/santilli_et_al_2005.pdf*.

11 Bill McKibben, in the *New York Review of Books* (16 November 2006) at *www.nybooks.com/articles/19596*.

Chapter 2.   **We're Not in Kansas Anymore**

12   Tornado in Dimmit, Texas: photograph by Harald Richter at *www.photolib.noaa.gov/nssl/nssl0179.htm*

14   Three hurricanes south of Japan on August 7, 2006, from *visibleearth.nasa.gov/view_rec.php?id=20946*, credit Jeff Schmaltz. "The slanting diagonal feature through the image is sunlight bouncing off the ocean into the MODIS instrument [on the satellite], a phenomenon called sunglint. The very bright swath is where the reflection is strongest."

14   Goran Ekstrom, Meredith Nettles, Victor C. Tsai, "Seasonality and increasing frequency of Greenland glacial earthquakes." *Science* 311 (2006): 1756–1758.

15   Jared Diamond, *Collapse: How Societies Choose to Fail or Succeed.* (Viking 2005), 6.

16   David Montgomery, *Dirt: The Erosion of Civilizations* (University of California Press, 2007).

17   Muir glacier pair: *nsidc.org/cgi-bin/gpd_run_pairs.pl*

19   Richard Lindzen, in *Newsweek* (2007) at *www.msnbc.msn.com/id /17997788/site/newsweek*. See also Daniel Grossman's interview, "Profile: Dissent in the Maelstrom," *Scientific American* (November 2001). Lindzen is a serious climate scientist who thinks that an "infrared Iris" associated with stratus cloud production and tall thunderheads will result in a climate sensitivity of only one-third the IPCC estimates. I hope he is right, though personally I would not go around telling people not to worry on the strength of a preliminary theory—nor describe ExxonMobil as "the only principled oil and gas company I know in the US." See *news.bbc.co.uk/2/low/business/ 6595369.stm*.

For the less established climate dissenters, a tendency to shift targets with time raises questions of whether it's really about the science or about something else. "Whatever the science is, they will try to find ways to question it," says Naomi Oreskes, a geologist and science historian at the University of California, San Diego. "That makes it clear that the issue for them is not the science." See Michael Hopkin, "Climate sceptics switch focus to economics: As the scientific case strengthens, dissenters change tack." *Nature* (10 February 2007) 582, at *dx.doi.org/10.1038/445582a*.

Floods and wildfires are from the figures in the 2007 IPCC report, in the WG 1 Summary for Policymakers at *www.ipcc.ch.* For zonal precipitation: Xuebin Zhang, et al, "Detection of human influence on twentieth-century precipitation trends," *Nature* 448(26 July 2007): 461–465, at *dx.doi.org/10.1038/nature06025.*

Governor Arnold Schwarzenegger of California, quoted by Thomas Friedman, "The power of green," *New York Times* Magazine (15 April 2007) at *www.nytimes.com/2007/04/15/magazine/15green.t.html.*

The correlation of temperature with western U.S. wildfires is from A. L. Westerling, H. G. Hidalgo, D. R. Cayan, and T. W. Swetnam. "Warming and earlier spring increase western U.S. forest wildfire activity." *Science* 313 (5789): 940 (18 August 2006). See *dx.doi.org/10.1126/science.1128834.*

19 John Holdren, "The energy innovation imperative." *Innovation* (Spring 2006): 11.

Chapter 3.   **Will This Overheated Frog Move?**

22 Svante Arrhenius, "On the influence of carbonic acid in the air upon the temperature of the ground." *Philosophical Magazine and Journal of Science (fifth series)* 41 (1896): 237–275. See Weart's history at *www.aip.org/history/climate.*

Roger Revelle and Hans E. Suess. "Carbon dioxide exchange between atmosphere and ocean and the question of an increase of CO_2 during the past decades." *Tellus* 9 (1957) 18-27.

26 Jim Hansen's presentation at the National Academy of Sciences in April 2006 is on his Columbia University website, *www.columbia.edu/~jeh1.*

28 Ozone hole, see *earthobservatory.nasa.gov/Newsroom/NewImages /images.php3?img_id=17436.*

29 *www.monbiot.com/archives/2005/10/25/our-own-nuclear-salesman/.*

30 Ten recommendations for reducing U.S. carbon emissions:

1. Immediately freeze carbon dioxide emissions and then begin a program to reduce them by at least 90% by 2050.

2. Replace the payroll tax for Social Security and Medicare with a tax on pollution, particularly carbon dioxide.

3. Use a portion of the tax on pollution to help low-income individuals adapt as carbon emissions are reduced.

4. Work towards de-facto compliance with the Kyoto Protocol to the United Nations Framework Convention on Climate Change, and create a new, strong international treaty with a starting date of 2010 instead of 2012.

5. Enact a moratorium on the construction of any new coal-fired power plants that are not compatible with carbon capture and sequestration.

6. Create an 'Electranet,' a smart grid in which power generation is widely distributed. Homeowners and small businesses could use solar and wind energy generators and sell that energy into the grid at a rate that is determined by the market.

7. Raise Corporate Average Fuel Economy (CAFE) standards for automobiles, and set energy standards for other industries.

8. Set a date for a ban on incandescent light bulbs.

9. Create a 'Connie Mae,' a carbon-neutral mortgage association that would help homebuyers pay for energy reduction measures such as insulation and energy-efficient windows that can have high upfront expenses.

10. Have the Securities and Exchange Commission (SEC) require the disclosure of carbon emissions in corporate reporting.

From The Honorable Al Gore's testimony to the U.S. Congress in March 2007. See *EOS* 88(10 April 2007): 171.

All necessary, but far too weak. Maybe this is what it takes to get Congress moving at last, but those ten are the easy stuff, what would have been appropriate twenty years ago. If we don't do considerably more, and quickly, it will be like rearranging the desk chairs on the *Titanic*.

I would instead emphasize the 2020 urgency requiring many new nuclear or geothermal power plants, retiring many old coal plants, converting to plug-in hybrid electric vehicles, and subsidizing DC power lines to developing countries with coal.

30   UN Secretary-General Ban Ki-moon, at press conference (8 June 2007). See *www.UN.org/apps/sg/offthecuff.asp?nid=1035*.

31   Andrew Revkin, "Global meltdown," *AARP Magazine* (July 2007) at *www.aarpmagazine.org/lifestyle/global_meltdown.html*.

32   R. A. Bindschadler et al, "Tidally controlled stick-slip discharge of a West Antarctic ice stream," *Science* 301 (2003):1087–89.

32 Joseph J. Romm, *Hell and High Water* (William Morrow, 2007), 2.

32 General Gordon R. Sullivan, *securityandclimate.cna.org/report* at p.10.

Chapter 4.   **"Pop!" Goes the Climate**

39 This satellite image shows the 2002 breakup of the Larsen B Ice Shelf. See Eugene Domack et al, "Stability of the Larsen B ice shelf on the Antarctic Peninsula during the Holocene epoch." *Nature* 436 (2005): 681–685. This region, covering approximately 3250 km² with 200 m thick ice, had been continuously glaciated since before the end of the last glacial period. Adapted from NASA Terra/MODIS imagery via *www.GlobalWarmingArt.com*. The 2005 melt/refreeze episode is at *earthobservatory.nasa.gov/Newsroom/NewImages/images.php3?img_id=176 61*

37 William H. Calvin, "The great climate flip-flop," *Atlantic Monthly* (January 1998). The story is elaborated in *A Brain for All Seasons: Human Evolution and Abrupt Climate Change.* (University of Chicago Press, 2002).

Chapter 5.   **Drought's Slippery Slope**

Error! Bookmark not defined. "Fleeing a dust storm" shows farmer Arthur Coble and his sons in Cimmaron County, Oklahoma. By Arthur Rothstein, U.S. government photographer, April, 1936 (Library of Congress LOC-00241v.jpg).

40 My synopsis of drought feedbacks derives from a brief talk in 2000 by J. M. Wallace.

43 Recent evaporation seeding the next rainfall: that is going to be a big problem in the Amazon. Today, the flat bottom of the clouds (where the dew point is) isn't very high off the ground. But with greenhouse warming, that flat bottom will move up to much higher in the sky—and so not mix very well with the recent evaporation. The clouds will continue westward until running into the Andes and dropping some rain there. It will flow down the Amazon river as it does now but the lush vegetation on the riverbanks will be gone— likely burned off during the onset of drought.

46 Connie A. Woodhouse, Jonathan T. Overpeck, "2,000 Years of Drought Variability in the Central United States," *Bulletin of the*

American Meteorological Society 79(1998): 2693–2714.

49   Worster quote at *www.pbs.org/wgbh/amex/dustbowl/peopleevents /pandeAMEX06.html.*

50   Also from Woodhouse and Overpeck (1998).

50   A 1 m rise in sea level would change the frequency of what are now 100-year floods in metropolitan New York to once in every four years events. See *ccir.ciesin.columbia.edu/nyc/ccir-ny_q2a.html* and C. Rosenzweig and W.D. Solecki (Eds.). *Climate Change and a Global City: The Potential Consequences of Climate Variability and Change - Metro East Coast.* Report for the U.S. Global Change Research Program, National Assessment of the Potential Consequences of Climate Variability and Change for the United States (Columbia Earth Institute, New York, 2001).

52   Although originally named the Medieval Warm Period, the temperature change does not seem to have been uniform around the globe. It is best thought of as a period of widespread climate anomalies preceding the better-defined Little Ice Age.

52   Drought frequency: Richard Seager et al, "Model projections of an imminent transition to a more arid climate in southwestern North America," *Science* 316 (25 May 2007): 1181–1184 at *dx.doi.org /10.1126/science.1139601.*

52   Eleanor J. Burke, Simon J. Brown, Nikolaos Christidis, "Modeling the Recent Evolution of Global Drought and Projections for the Twenty-First Century with the Hadley Centre Climate Model," *Journal of Hydrometeorology* 7 (October 2006): 1113–1125.

53   Dust wall photos from NOAA's George E. Marsh Album via *commons.wikimedia.org/wiki/Image:Dust.*

57   Irrigation, see *ga.water.usgs.gov/edu/irsprayhigh.html* and *www.worldwatch.org/node/811.*

58   Kansas crop circle irrigation photo from *earthobservatory.nasa.gov /Newsroom/NewImages/Images/kansas_AST_2001175_lrg.jpg.*

Chapter 6   **Why Deserts Expand**

59   Kim Stanley Robinson, "Imagining Abrupt Climate Change: Terraforming Earth." Amazon Short essay (2005).

61   George Hadley, "Concerning the cause of the general trade winds," *Philosophical Transactions*, 39 (1735).

62 Qiang Fu, Celeste M. Johanson, John M. Wallace, Thomas Reichler. "Enhanced mid-latitude tropospheric warming in satellite measurements," *Science* 312 (26 May 2006): 1179.

62 Richard Seager, et al., "Model projections of an imminent transition to a more arid climate in southwestern North America." *Science* 316 (25 May 2007): 1181–1184 at *dx.doi.org/10.1126/science.1139601*.

62 Perth: Flannery (2005): 130. Pittock (2005): 141.

63 Long-term drought indicator blends at *www.drought.unl.edu/dm/monitor.html*.

68 Daniel Nepstad, see *www.whrc.org/resources/online_publications/essays/2006-08-Nepstad-Independent.htm*.

68 George Monbiot, *Heat* (2006): xi.

68 Fred Pearce, "Global meltdown." *The Guardian* (30 August 2006). *environment.guardian.co.uk/climatechange/story/0,,1860560,00.html*

Chapter 7.   From Creeps to Leaps

70 U.S. Coast Guard photograph of New Orleans on the day after Hurricane Katrina, 2005, after three levees had failed.

72 Robert Frost, in *Selected Prose of Robert Frost*, edited by H. Cox and E. C. Lathem (Collier, 1986), 33–46.

73 Snowballing, see Pittock (2005): 110ff, for an excellent discussion on nonlinear effects in climate.

75 The Teton Dam, 44 miles northeast of Idaho Falls in southeastern Idaho, failed abruptly on June 5, 1976 when being filled for the first time. Engineers were actively looking for leaks and saw a wet spot. However, the collapse progressed so rapidly that several large bulldozers were lost and downstream communities only had one hour of warning. The dam failure released nearly 300,000 acre feet of water, which flooded farmland and towns downstream at the loss of 14 lives and a cost of $1 billion. See *npdp.stanford.edu/npdphome/npdpimages/Photo%20Gallery/fullimages/IDS00007_003_f.jpg*.

76 Thomas R. Malthus, *An Essay on the Principle of Population* (printed for J. Johnson, London, 1798).

77 Restaurant lead-lag dynamics: *muller.lbl.gov/pages/news%20reports/ebexp.htm*.

81   David Montgomery, *Dirt: The Erosion of Civilizations* (University of California Press, Berkeley, 2007).

82   James Martin, *The Meaning of the 21st Century* (Eden Project, London, 2007): 30. Graph of cod catch from *www.fao.org/docrep/005 /y3684e/y3684e05.htm*.

85   Al Gore, on the Charlie Rose Show (September 2006).

85   Raymond T. Pierrehumbert, "Climate Change: A Catastrophe in Slow Motion," *Chicago Journal of International Law* 6 (Winter 2006): 573. See *cjil.uchicago.edu/past-issues/win06.html*.

Chapter 8.   **What Makes a Cycle Vicious?**

88   Watt story from James Lovelock, *The Revenge of Gaia* (Allen Lane: London, 2006). See *en.wikipedia.org/wiki/Centrifugal_governor*. The figure is from "Discoveries & Inventions of the Nineteenth Century" by R. Routledge, 13th edition, published 1900.

89   Good use is made of positive feedback by nerve cells and muscles. It's what makes things happen quickly. It shortens your reaction time enough so that you have quick reflexes and don't go bouncing down a flight of steps. Your computer uses positive feedback in much the same way to shorten each step of the computing cycle. When the first flip-flop circuits were invented for computer bits, they operated on about the same timescale as nerve cells (milliseconds). Now they (but not the nerve cells) are a million times faster, operating in nanoseconds.

92   James Hansen, Makiko Sato, Pushker Kharecha, Gary Russell, David W Lea, Mark Siddall, "Climate change and trace gases." *Philosophical Transactions of the Royal Society A* (July 15, 2007) at *dx.doi.org/10.1098/rsta.2007.2052*.

94   K. Steffen, R. Huff, *Greenland Melt Extent, 2005* (Cooperative Institute for Research in Environmental Sciences, University of Colorado, Boulder, 2005); available at *cires.colorado.edu/science /groups/steffen/greenland/melt2005/*.

95   Water vapor amplifies CO_2 warming by 40–50 percent: IPCC 2007 WG1 SPM.

96   "When CO_2 increases the storage of heat in the lower atmosphere, it promotes more evaporation from the tropical oceans." By itself, this positive feedback is somewhat self-regulating as high

humidity means more clouds and their whiteness reflects some
sunlight back out into space, somewhat countering the heating effect
of more water vapor in the atmosphere. Since cloud formation also
depends on a number of other things such as the size of smoke
particles, the balancing act is not well understood yet. For example,
agricultural fires create soot whose particles are large enough to seed
water droplet formation. Power plants burning fossil fuels produce
smaller particles and less rainfall downwind.

96   Frank J. Wentz, Lucrezia Ricciardulli, Kyle Hilburn, Carl Mears,
"How Much More Rain Will Global Warming Bring?" *Science* (31 May
2007) at *dx.doi.org/10.1126/science.1140746.*

98   Ted Scambos quote at *earthobservatory.nasa.gov/Newsroom/
MediaAlerts/2006/2006100323310.html.*

99   Eleanor J. Burke, Simon J. Brown, Nikolaos Christidis,
"Modeling the Recent Evolution of Global Drought and Projections
for the Twenty-First Century with the Hadley Centre Climate
Model," *Journal of Hydrometeorology* 7 (October 2006): 1113–1125.

Chapter 9.   **That Pale Blue Sky**

101   David Appell, news article "The darkening Earth," *Scientific
American* (August 2004): 16–17.

In addition to balloons, sulfur could be distributed via jet fuel.
To avoid adding sulfur to the lower atmosphere, one fuel tank on an
airliner would be filled with sulfur-free fuel and used on the climb up
to cruising altitude (which accounts for about one-quarter of a long
flight's fuel consumption). But when cruising above the weather, the
sulfur-enhanced jet fuel would be used.

106   V. Ramanathan, quoted by Pearce (2006).

107   Richard A. Feely, Christopher L. Sabine, and Victoria J. Fabry,
"Carbon dioxide and our ocean legacy." National Environment Trust
brochure (2006), see *www.NET.org/documents/ocean_acidification.*

107   Oliver Morton, "Is this what it takes to save the world?" *Nature*
447 (10 May 2007):132-136 news feature at *dx.doi.org/10.1038/447132a.*

107   El Niño reduces primary production: Michael J. Behrenfeld, et
al, "Climate-driven trends in contemporary ocean productivity,"
Nature 444 (7 Dec 2006): 752–755. At *dx.doi.org/10.1038/nature05317.*

107   Power plant fallout map adapted from exhibit 3–1 in *cta.policy.net/fact/mortality/mortalityabt.pdf*

108   Tim Flannery (2006): 219.

109   Fire maps created by Jacques Descloitres, MODIS Rapid
Response System at NASA/GSFC. See *rapidfire.sci.gsfc.nasa.gov
/firemaps/*.

110   Deaths from coal, adapted from Exhibit 6–1 Premature
Mortality Risk Attributable to PM2.5 from Power Plants, 2010
Baseline, at *www.cleartheair.org/dirtypower/docs/abt_powerplant_whitepaper.pdf*

Chapter 10.   **Slip Locally, Crash Globally**

111   Richard Alley, quoted by Pearce (2006).

112   The Google Earth software is at *earth.google.com*; once installed,
go to *WilliamCalvin.com/2006/GoogleEarth_PlacemarkGreenland
Shoulder.kmz* for a view of the pockmarked western shoulder of
Greenland. Startup and find the terrain toggle so there is a readout of
Lat/Long/Elev. Once positioned over the west coast of Greenland at
about 70°N, start moving south, zooming in on the long east-west
tongue of Jakobshaven Isbrae (once an ice shelf, until warmer waters
undermined it and broke it up like Larsen B). Then move east to see
the lakes on the shoulder of the ice sheet. Finally travel south,
keeping lakes in sight. The drainage of these lakes is likely setting up
the collapse of the southern half of the central Greenland ice sheet.

113   Melt water lake photograph is by climate scientist Sarah Das.
www.whoi.edu/oceanus/viewImage.do?id=17710&aid=9126. See NASA
photos at *earthobservatory.nasa.gov/Newsroom/NewImages
/images.php3?img_id=17607*.

114   Greasing the skids: Jay Zwally et al, "Surface melt-induced
acceleration of Greenland ice-sheet flow," *Science* 297(2002): 218–222.
Also, from studies in Iceland, the water that gets trapped under the
ice cannot refreeze if it is under so much pressure that it cannot
expand into ice. And so it is forced up into whatever cracks the icy
bottom affords. If finding space to expand, it freezes. The heat given
up in freezing warms the surrounding ice, beginning a self-
destructive cycle along the bottom of the ice sheet that crumbles the
attachment to the bedrock. There's more at *www.pbs.org/wgbh/nova
/transcripts/3211_megafloo.html*.

115   Quoted by Pearce (2006): 70.

119   Sea-level rise, modified from figure 5.5.2 of the 2007 IPCC report SPM. It comes from an oxygen isotope record for the Red Sea over the past 470 kyr. See Siddall et al., *Nature* 423 (2003): 853–858.

119   Plankton appear in various roles in scenarios for pumping down carbon in an ice age. Fertilization: the higher winds of an ice age should carry a lot more iron-rich dust into the Atlantic from the Sahara and Namib deserts. Expanded habitat: the reduced meridional heat transport during an ice age cools the North Atlantic, and Lovelock (2006) argues for the cooler oceans allowing plankton to thrive in more places and so pumping down carbon faster. The jury is still out on their relative importance, and certainly regarding how they might be manipulated to solve our CO_2 problem.

120   J. W. Day, Jr., et al, "Emergence of complex societies after sea level stabilized." *EOS* 88 (10 April 2007): 169.

122   Richard Alley quote from Pearce (2006).

122   David D. Zhang, Jane Zhang, Harry F. Lee, Yuan-qing He. "Climate Change and War Frequency in Eastern China over the Last Millennium." *Human Ecology* 35 (2007):403–414 at *dx.doi.org/10.1007/s10745-007-9115-8*.

127   Flannery (2006): 140.

Chapter 11.   **Come Hell and High Water**

131   Most of the sea-level maps were produced, thanks to Jonathan Overpeck and Jeremy Weiss, with the mapping software at the University of Arizona. See *www.geo.arizona.edu/dgesl/research/other /climate_change_and_sea_level/sea_level_rise/sea_level_rise.htm*.

133   Max Mayfield, director of the National Hurricane Center, quoted in Reuters interview (22 August 2006).

134   NYC storm surge proposal, see *www.nyas.org/ebriefreps /main.asp?intSubsectionID=2247* and *metroeast_climate.ciesin.- columbia.edu/reports/infrastructure.pdf*. For more on storm surges, with New York City and Long Island examples, see *stormy.msrc.sunysb.edu*. Includes real-time forecasts.

 135   PATH station in Hoboken during a 1992 nor'easter. This and the WTC entry photos are from the Metro New York Hurricane Transportation Study, 1995.

Sir David King, "Climate Change Science: Adapt, Mitigate, or

Ignore?" *Science* 303 (9 January 2004): 176–177.

141   Claudia Dreifus, "A Conversation With Jeffrey Mount: Giving Sacramento Good Reason to Have New Orleans on Its Mind." *New York Times* (April 18, 2006). *www.nytimes.com/2006/04/18/science/18conv.html.*

142   David Biello, "Conservative Climate: Consensus document may understate the climate change problem," On the Scientific American web site (18 March 2007) at *www.sciam.com/print_version.cfm?-articleID=5B9E73AD-E7F2-99DF-3F71280BCE41ED77.*

145   The graph of sea-level rise in the last 24,000 years is from the 2007 IPCC Summary for Policymakers WG1. I have extensively modified the Washington Monument photograph at *en.wikipedia.org-/wiki/Image:Washington_Monument_Dusk_Jan_2006.jpg.*

My print-your-own color poster, suitable for posting within 20 ft of sea level. is in a PDF file at *Global-Fever.org/#posters*, together with directions for determining where to post it using GPS units or Google Earth. For a similar project, see *nytimes.com/2007/06/16/arts/design/16chal.html.*

146 Göran Ekström, Meredith Nettles, and Victor C. Tsai, "Seasonality and Increasing Frequency of Greenland Glacial Earthquakes," *Science* (24 March 2006): 1756–1758 at *dx.doi.org/10.1126/science.1122112.*

146 Jonathan T. Overpeck, Bette L. Otto-Bliesner, Gifford H. Miller, Daniel R. Muhs, Richard B. Alley, and Jeffrey T. Kiehl, "Paleoclimatic evidence for future ice-sheet instability and rapid sea-level rise," *Science* (24 March 2006)1747–1750 at *dx.doi.org/10.1126/science.1115159.*

146 Eric Rignot and Pannir Kanagaratnam, "Changes in the velocity structure of the Greenland Ice Sheet," *Science* (17 February 2006): 986–990. At *dx.doi.org/10.1126/science.1121381.*

147 For a discussion of the 2007 IPCC estimates of sea-level rise, see Stefan Rahmstorf's discussion at *www.realclimate.org/index.php /archives/2007/03/the-ipcc-sea-level-numbers/#more-427.*

148 Mark Lynas, *Six Degrees* (Fourth Estate, 2007): 171.

149 Modeling data from Kurt M. Cuffey and Shawn J. Marshall, "Substantial contribution to sea-level rise during the last interglacial from the Greenland ice sheet," *Nature* 404 (6 April 2000): 591-594. At *dx.doi.org/10.1038/35007053.*

149 Greenland ice about 125,000 years ago is inferred from models; see 2007 IPCC report WG1 technical chapters.

Chapter 12.   **Methane Is the Double Threat**

150 CO_2, CH_4 and temperature records from James E. Hansen, *Climate Change* 68 (2005): 269.

151 Kirpotkin, quoted by Pearce (2006): 111.

152 When you hear the phrase, "Doubling CO_2," it refers to the pre-industrial CO_2 level of 275 ppm being doubled to 550 ppm of CO_2 equivalents. The natural range for CO_2 between ice-age minima and maxima is about 100 parts per million. We have already gone 110 ppm past the historical maximum and need to add at least 50 ppm for the CO_2 equivalents of the increased concentrations of methane and other GHGs.

152 Guiseppe Etiope, "The geological links of the ancient Delphic Oracle (Greece): A reappraisal of natural gas occurrence and origin." *Geology* (October 2006): 825–828. Since methane dilutes the oxygen concentration of the air, anoxia likely affected the Oracle.

152   Renato Spahni, Jérôme Chappellaz, Thomas F. Stocker, Laetitia
Loulergue, Gregor Hausammann, Kenji Kawamura, Jacqueline
Flückiger, Jakob Schwander, Dominique Raynaud, Valérie Masson-
Delmotte, and Jean Jouzel. "Atmospheric Methane and Nitrous Oxide
of the Late Pleistocene from Antarctic Ice Cores," *Science* (25
November 2005): 1317–1321.

152   John M. Barry, *Rising Tide: The Great Mississippi Flood of 1927 and
How It Changed America* (Simon and Shuster, 1997) 69.

154   Natural gas leaks about 2 to 4 percent: Society of Chemical
Industry (2004), quoted in Lovelock (2006) p.75.

David A. Kirchgessner, Robert A. Lott, R. Michael Cowgill,
Matthew R. Harrison, Theresa M. Shires, "Estimate of methane
emissions from the U.S. natural gas industry," n.d., at
www.epa.gov/ttn/chief/ap42/ch14/related/methane.pdf.

154   LNG emissions: Richard Heede, May 2006 report at
www.edcnet.org/ProgramsPages/LNGrptplusMay06.pdf.

154   T. M. Hill et al, "Climatically driven emissions of hydrocarbons
from marine sediments during deglaciation," *Proceedings of the
National Academy of Sciences (U.S.)* 103 (12 September 2006):13570 at
dx.doi.org/10.1073/pnas.0601304103.

A. I. Best et al, "Shallow seabed methane gas could pose coastal
hazard," *Eos* 87 (30 May 2006): 1.

Aerial photo by Katey Walter of Siberian thaw lakes near
Cherskii, Siberia in 2003. Siberian methane video at
*mms://ms.groovygecko.net/groovyg/clients/nmsi/scim/antenna/siberianthaw
/Siberian_broadband.wmv.*

Katey Walter et al, "Methane bubbling from Siberian thaw lakes
as a positive feedback to climate warming." *Nature* (7 September
2006): 443, at *dx.doi.org/10.1038/nature05040.*

David Archer, "Methane hydrates and anthropogenic climate
change." *Biosci. Discuss.* 4 (2007): 993–1057 and see *RealClimate.org/
index.php?p=227.*

Burning methane hydrate photograph from *www.giss.nasa.gov/
research/features/methane/hydrate.jpg.*

157   I. J. Simpson, F. S. Rowland, S. Meinardi, and D. R. Blake,
"Influence of biomass burning during recent fluctuations in the slow
growth of global tropospheric methane," *Geophysiocal Research Letters*

33 (2006): L22808, *dx.doi.org/10.1029/2006GL027330.*

157 David Archer and Victor Brovkin, "Millennial Atmospheric Lifetime of Anthropogenic CO2," *Climatic Change* (to appear, 2007).

158 Ocean acidification figure adapted from the Hadley Centre's HadOCC model; via John Holdren's MBL slide (2006).

159 Marten Scheffer, Victor Brovkin and Peter Cox, "Positive feedback between global warming and atmospheric CO2 concentration inferred from past climate change." *Geophysical Research Letters* (2006) at *dx.doi.org/10.1029/2005GL025044.*

159 M. S. Torn and J. Harte, "Missing feedbacks, asymmetric uncertainties, and the underestimation of future warming," *Geophysical Research Letters* (2006) at *dx.doi.org/10.1029/2005GL025540.*

159 Agricultural waste problem, see *www.virtualcentre.org/en /library/key_pub/longshad/A0701E00.pdf.*

Chapter 13.   **Sudden Shifts in Climate**

162 "Winds gusting to more than 100 mph swept across northern Utah on Friday, overturning 20 tractor-trailers....Winds reached 113 mph setting a state record...." Photo by Marta Storwick for the *Standard-Examiner* of Ogden, Utah (23 April 1999), with permission.

163 Claudia Tebaldi et al, "Going to Extremes," *Climatic Change* (December 2006). See *www.ucar.edu/news/releases/2006 /wetterworld.shtml.*

Richard B. Alley, et al., "Abrupt climate change." *Science* 299 (2003): 2005–10.

164 See "How Likely are Major or Abrupt Climate Changes, such as Loss of Ice Sheets or Changes in Global Ocean Circulation?"in G. A. Meehl, et al, "Global Climate Projections. Section 10.2 in *Climate Change 2007: The Physical Science Basis. Contribution of Working Group I to the Fourth Assessment Report of the Intergovernmental Panel on Climate Change,* edited by Susan Solomon, et al. (Cambridge University Press, Cambridge and New York) at *ipcc-wg1.ucar.edu/wg1/Report /AR4WG1_10.pdf.*

164 William H. Calvin, "The great climate flip-flop," *Atlantic Monthly* (January 1998). The story is elaborated in *A Brain for All Seasons: Human Evolution and Abrupt Climate Change.* (University of Chicago Press, 2002).

166   Illustration adapted from the National Oceanic and
Atmospheric Administration's El Niño Web site, *www.pmel.noaa.gov
/tao/elnino*. Technically, an El Niño is when mid-Pacific sea surface
temperature stays more than 0.5°C above normal for four months. A
La Niña is when it cools more than 0.5°C for four months (although
some may use La Niña for the normal midrange as well). A La Niña
situation often follows an El Niño episode and is essentially its
opposite. During a La Niña, the easterly trade winds near the equator
are stronger than normal. They push more warm surface waters
westward across the Pacific. The colder, deeper waters that well up to
the surface in their place extend far out into the central equatorial
Pacific. The historical El Niño chart is at *www.cpc.ncep.noaa.gov/
products/analysis_monitoring/lanina/enso_evolution-status-fcsts-web.ppt*

168   A. V. Fedorov et al, "The Pliocene paradox (mechanisms for a
permanent El Niño)," *Science* 312 (9 June 2006): 1485 at *dx.doi.org
/10.1126/science.1122666*.

Chapter 14.   **A Sea of CO₂**

172   The data snapshot is from March to June, the northern sunlight
making the bloom there more than in the southern hemisphere
winter. Nutrients are a major limitation. Besides nutrients from rivers,
they are also up-welled to the surface in some areas (line in mid-
Pacific Ocean where trade winds converge, also on the west coast of
continents). For the original color version of the world phytoplankton
imaging, see *earthobservatory.nasa.gov/Newsroom/NewImages/
images.php3?img_id=17332*. Maps, see *neo.sci.gsfc.nasa.gov/
Search.html?group=12*. If you have Google Earth installed, see
neo.sci.gsfc.nasa.gov/RenderData?si=493385&cs=rgb&format=KMZ.

   The pictures are thanks to Russell Hopcroft (*Cavolinia uncinata*,
left), Victoria Fabry (*C. tridentata*, right), and Laurence Madin (*Salpa
aspera*).

173   B. Schmitz, "Plankton cooled a greenhouse." *Nature* (14
September 2000) 407: 143–144.   `

174   Michael P. Lesser, "Coral reef bleaching and global climate
change: Can corals survive the next century?" *Proceedings of the
National Academy of Sciences (U.S.)* (2007) at *dx.doi.org/10.1073-
/pnas.0700910104*.

174   Caribbean coral losses in 2005, see *news.nationalgeographic.com*

/news/2006/04/0406_060406_coral.html

175   J. F. Bruno, et al., "Thermal stress and coral cover as drivers of coral disease outbreaks." *PLoS Biology* 5 (2007): e124 at *dx.doi.org /10.1371/journal.pbio.0050124.*

175   Policy statement from the Royal Society, London, "Ocean Acidification Due to Increasing Atmospheric Carbon Dioxide" (2005) at *www.royalsoc.ac.uk/displaypagedoc.asp?id=13539.*

175   J.A. Kleypas, R.A. Feely, V.J. Fabry, C. Langdon, C.L. Sabine, and L.L. Robbins. "Impacts of Ocean Acidification on Coral Reefs and Other Marine Calcifiers: A Guide for Future Research" 2006 report of a workshop held 18–20 April 2005, St. Petersburg, FL, 88 pp, at *www.ucar.edu/communications/Final_acidification.*

176   Corinne Le Quéré, et al. "Saturation of the Southern Ocean CO_2 Sink Due to Recent Climate Change," *Science* (2007) at *dx.doi.org /10.1126/science.1136188.*

179   Plankton decline: Scott C. Doney, "Plankton in a warmer world." *Nature* 444 (7 December 2006) 695.

179   Michael J. Behrenfeld, et al. "Climate-driven trends in contemporary ocean productivity." *Nature* 444 (7 December 2006): 752–755 at *dx.doi.org/ 10.1038/nature05317.*

179   Ken Calderia, quoted by Elizabeth Kolbert, "The Darkening Sea: What carbon emissions are doing to the ocean,"*New Yorker* (20 November 2006): 67–75.

181   Philip W. Boyd, et al. "Mesoscale Iron Enrichment Experiments 1993–2005: Synthesis and Future Directions." *Science* 315 (2 February 2007): 612–617 at *dx.doi.org/10.1126/science.1131669.*

181   Philip W. Boyd, "Biogeochemistry: Iron findings." *Nature* 446 (26 April 2007): 989–991.

181   Kerguelen analysis: Stèphane Blain, et al. "Effect of natural iron fertilization on carbon sequestration in the Southern Ocean." *Nature* 446 (26 April 2007): 1070–1074 at *dx.doi.org/10.1038/nature05700.* Kerguelen figure modified from the accompanying news article.

184   The wave-driven pump to raise deep water to the surface is best seen in the archived presentations at *atmocean.com.*

Chapter 15.   **The Extended Forecast**

187   Joseph J. Romm, *Hell and High Water* (William Morrow, 2007): 8.

187   Mark Lynas, *Six Degrees* (Fourth Estate, London, 2007): 171.

188   The 2020 turnaround is modified from a slide in John Holdren's MBL talk (November 2006), see *www.whrc.org/resources/PPT /JPH_MBL_11-03-06_Clim-Chg-Challenge.ppt.*

188   Tony Blair, quoted by Flannery (2005): 247.

190   The photograph shows a 1 meter section of the GISP2 ice core from a depth of 1837 meters in the Greenland Ice Sheet. From *GlobalWarmingArt.com/wiki/Image:GISP2_Ice_Core_jpg.*

192   Susan Solomon, D. Qin, M. Manning, R.B. Alley, T. Berntsen, N.L. Bindoff, Z. Chen, A. Chidthaisong, J.M. Gregory, G.C. Hegerl, M. Heimann, B. Hewitson, B.J. Hoskins, F. Joos, J. Jouzel, V. Kattsov, U. Lohmann, T. Matsuno, M. Molina, N. Nicholls, J. Overpeck, G. Raga, V. Ramaswamy, J. Ren, M. Rusticucci, R. Somerville, T.F. Stocker, P. Whetton, R.A. Wood and D. Wratt, "Technical Summary. In: *Climate Change 2007: The Physical Science Basis. Contribution of Working Group I to the Fourth Assessment Report of the Intergovernmental Panel on Climate Change* [Susan Solomon, D. Qin, M. Manning, Z. Chen, M. Marquis, K.B. Averyt, M. Tignor and H.L. Miller (eds.)]. Cambridge University Press, Cambridge, United Kingdom and New York, NY, USA. at *ipcc-wg1.ucar.edu/wg1/Report/AR4WG1_TS.pdf.* Summary for policymakers is at *www.ipcc.ch/SPM2feb07.pdf.*

198   Spencer Weart, *New Scientist* (14 April 2007) 20.

199   Stefan Rahmstorf, et al., "Recent climate observations compared to projections," *Science* 316 (4 May 2007): 709 at *dx.doi.org/10.1126/ science.1136843.*

199   James E. Hansen, "Scientific reticence and sea level rise." *Environ. Res. Lett.,* 2 (2007): 024002, at *dx.doi.org/10.1088/1748- 9326/2/2/024002.*

199   Mark Bowen, *Thin Ice* (Henry Holt, 2005).

199   Richard A. Kerr, "Pushing the Scary Side of Global Warming," *Science* 316(8 June 2007): 1412–1415, at *dx/doi.org/10.1126/ science.316.5830.1412.*

201   Wind turbine in Skåne, Sweden. Photograph by Väsk at *commons.wikimedia.org/wiki/Image:Vindkraftverk_i_Sk%C3%A5ne_februa ri_2005.jpg.*

202 Sir Crispin Tickell (2002), at *www.futurefoundation.org/documents/nty_report_apr02.pdf*

Chapter 16.   Doing Things Differently
204 Phillip F. Schewe, *The Grid* (Joseph Henry Press DC, 2007): 168.
206 California vs. US electrical use per person from *www.eia.doe.gov/emeu/states/sep_use/total/use_csv*. Table of all states at *www.eia.doe.gov/emeu/states/sep_sum/plain_html/rank_use_per_cap.html*.
207 Paul Krugman, "Colorless green ideas." *New York Times* (23 February 2007), *select.nytimes.com/2007/02/23/opinion/23krugman.html*.
209 Evolution of electricity sources from *www.iea.org/Textbase/stats/graphsearch.asp*.
217 Phillip F. Schewe, *The Grid* (Joseph Henry Press, 2006).

Chapter 17.   Cleaning Up Our Act
221 George Monbiot, *Heat* (2006): 44.
221 Paul Falkowski, et al., "The Global Carbon Cycle: A Test of Our Knowledge of Earth as a System." *Science* 290 (13 October 2000): 291. At *dx.doi.org/10.1126/science.290.5490.291*.
222 Wedges, see *www.princeton.edu/~cmi/resources-/stabwedge.htm*.
224 UN's 2007 expert group: *www.UNfoundation.org/SEG/*.

Chapter 18.   The Climate Optimist
228 Gregory L. Armstrong, Laura A. Conn, Robert W. Pinner. "Trends in Infectious Disease Mortality in the United States during the 20th Century." *JAMA* 281 (January 1999): 61–66.
231 E. O. Wilson (personal communication, 2006) put the extra-long, extra-warm El Niño time frame this way: "...could burn down so much of the remaining rain forests in Southeast Asia and the Amazon that as many as half the remaining species of plants and animals could face early extinction."
232 High-speed toll gates: A quick method that avoids the high costs of creating a new roadside network would be to install old cell-phone technology in the vehicle and simply use it to detect when the vehicle crosses from one cell to another—and billing accordingly. See *www.newscientisttech.com/channel/tech/mg19225815.600-cellphone-networks-could-help-with-road-tolls.html*.

232   In the U.S., many people pay far more in payroll taxes (mostly Social Security, Medicare, and unemployment taxes) than they have withheld for income tax.

235   Artificial photosynthesis, see Frédéric Goettmann, Arne Thomas, Markus Antonietti, "Metal-Free Activation of CO_2 by Mesoporous Graphitic Carbon Nitride." *Angewandte Chemie* (2007) at *dx.doi.org/10.1002/anie.200603478.*

236   FDR 1940–1941 leadership, see pp. 44–59 in Doris Kearns Goodwin's *No ordinary time* (Simon and Shuster, 1994).

237   Jack Doyle, *Taken for a ride: Detroit's Big Three and the politics of air pollution* (2000). Indeed, I'd say that Detroit's automakers may need a new purpose in life (and I'd suggest temporarily repurposing the manned part of NASA's space program as well). All of that talent is badly needed for more important tasks.

237   Gregg Easterbrook, "Some convenient truths," *Atlantic Monthly* (September 2006). *www.theatlantic.com/doc/print/200609/global-warming.*

Chapter 19.   **Turnaround by 2020**

238   James Hansen, "How Can We Avert Dangerous Climate Change?" Testimony before U.S. Congress (2007) at *arxiv.org/abs/0706.3720.*

239   M. Meinshausen "What does a 2°C target mean for greenhouse gas concentrations? A brief analysis based on multi-gas emission pathways and several climate sensitivity uncertainty estimates," in *Avoiding Dangerous Climate Change*, H. J. Schellnhuber, W. Cramer, N. Nakicenovic, T. Wigley, G. Yohe eds., (Cambridge University Press, Cambridge, 2006) at *www.cgd.ucar.edu/~mmalte/simcap /publications/meinshausenm_risk_of_overshooting_final_webversion.pdf.*

240   George Mueller (pronounced Miller), see *en.wikipedia.org/wiki /George_Mueller* and *futurefoundation.org/programs/ hum_wrk4.htm.*

243   Michael Kintner-Meyer, Kevin Schneider, Robert Pratt. "Impacts assessment of plug-in hybrid vehicles on electric utilities and regional U.S. power grids. Part 1: technical analysis." Report from U.S. Department of Energy's Pacific Northwest National Laboratory (2007) at *www.pnl.gov/energy/eed/etd/pdfs/*

phev_feasibility_analysis_combined.pdf

244 Compressed air car: see *en.wikipedia.org/wiki/Air_car* and "World's First Air-Powered Car: Zero Emissions by Next Summer" in *Popular Mechanics* (June 2007) at *www.popularmechanics.com/automotive/new_cars/4217016.html*. "India's largest automaker is set to start producing the world's first commercial air-powered vehicle. [It] can hit 68 mph and has a range of 125 miles. It will take only a few minutes for the CityCAT to refuel at gas stations equipped with custom air compressor units; MDI says it should cost around $2 to fill the car's carbon-fiber tanks with 340 liters of air at 4350 psi. Drivers also will be able to plug into the electrical grid and use the car's built-in compressor to refill the tanks in about 4 hours."

246 Zimmer Power Station photograph and data thanks to George Campbell. See *tallgeorge.com/Zimmer.htm*. Platt's provides an excellent selection of free maps for energy resources at *www.platts.com/Resources/map/archive/map_archive.html*.

246 For the nuclear fuels in the fly ash, see Alex Gabbard's analysis at *www.ornl.gov/info/ornlreview/rev26-34/text/colmain.html*.

246 Capture CO_2, see "Future of 'Clean Coal' Power Tied to (Uncertain) Success of Carbon Capture and Storage" at *www.sciam.com*.

246 Jeff Goodell, *Big Coal* (Houghton Mifflin, 2006).

246 MIT report, "The future of coal: options for a carbon-constrained world" (2007) at *web.mit.edu/coal/The_Future_of_Coal.pdf*.

From its summary illustrating the challenge of scale for carbon capture and long-term storage:

- Today fossil sources account for 80% of energy demand: Coal (25%), natural gas (21%), petroleum (34%), nuclear (6.5%), hydro (2.2%), and biomass and waste (11%). Only 0.4% of global energy demand is met by geothermal, solar and wind.
- 50% of the electricity generated in the U.S. is from coal.
- There are the equivalent of more than five hundred, 500 megawatt, coal-fired power plants in the United States with an average age of 35 years.
- China is currently constructing the equivalent of two, 500 megawatt, coal-fired power plants per week and a capacity comparable to the entire UK power grid each year.

- One 500 megawatt coal-fired power plant produces approximately 3 million tons/year of carbon dioxide (CO_2).
- The United States produces about 1.5 billion tons per year of CO_2 from coal-burning power plants.
- If all of this CO_2 is transported for sequestration, the quantity is equivalent to three times the weight and, under typical operating conditions, one-third of the annual volume of natural gas transported by the U.S. gas pipeline system.
- If 60% of the CO_2 produced from U.S. coal-based power generation were to be captured and compressed to a liquid for geologic sequestration, its volume would about equal the total U.S. oil consumption of 20 million barrels per day.
- At present the largest sequestration project is injecting one million tons/year of carbon dioxide (CO_2) from the Sleipner gas field into a saline aquifer under the North Sea.

254   MIT also assembled a panel of 18 experts in 2006 to evaluate large-scale use of deep geothermal, led by Jefferson W. Tester, the H.P. Meissner Professor of Chemical Engineering. The report, entitled "The Future of Geothermal Energy: Impact of Enhanced Geothermal Systems (EGS) on the United States in the 21st Century," is at *geothermal.inel.gov/publications/future_of_geothermal_energy.pdf*.

254   The Hot Dry Rock concept at the Los Alamos National Laboratory in 1972, see *www-geo.lanl.gov/expertise/geotherm.htm*. R. M. Potter, E. S. Robinson, and M. C. Smith, "Method of extracting heat from dry geothermal reservoirs," U.S. Patent 3,786,858 (1974).

Dave Duchane and Don Brown, "Hot Dry Rock (HDR) geothermal energy research and development at Fenton Hill, New Mexico," *GHC Bulletin* (December 2002): 13–19, at *geoheat.oit.edu-/bulletin/bull23-4/art4.pdf*. "It was found entirely feasible to operate the plant for extended periods of time with no on-site personnel, a fact that has important economic implications for the ultimate commercialization of HDR technology." This refers to the recirculating well side of the system with heat exchanger, not a complete plant with subsequent electricity generation from the heat exchanger. From "Building a Hot Rock Energy System" at *hotrock.anu.edu.au*:

Heat is extracted by pumping water through an engineered heat exchanger connecting two or more wells. This heat exchanger is a volume of hot dry rock with enhanced permeability. It is fabricated by hydraulic stimulation. This involves pumping high pressure water into the pre-existing fracture system that is present in all rocks to varying degrees. The high pressure water opens the stressed natural fractures and facilitates micro-slippage along them. When the water pressure is released, the fractures close once more but the slippage that occurred prevents them from mating perfectly again. The result is a million-fold permanent increase in permeability along the fracture systems and a heat exchanger that can be used to extract energy.

In a typical system, an initial borehole is sunk into the hot rock mass and a hydraulic stimulation is performed. A three dimensional microseismic network deployed on the surface and in nearby wells is used to record the little noises caused by the fractures widening as the pumping continues over several weeks. In this way, the progress of the stimulation is monitored and the size and shape of the growing heat exchanger is mapped.

A second well is then drilled into the margin of the heat exchanger 500 m or more from the first well. Now water can be pumped through the underground heat exchanger and in super-heated form it can be returned to the surface. There it can have its energy extracted before being reinjected to go around the loop again.

Of course, if you drill in an earthquake-prone area, the little earthquakes that result may become strong enough to feel (see *en.wikipedia.org/wiki/Hot_dry_rock_geothermal*). If there are also a lot of people around to notice—as was the case in a Basel suburb in December 2006—much fuss may result even for 3.4 strength earthquakes with no injuries. The reason for locating the wells within the city was presumably the 2,700 households to be heated from the plant's excess (in addition to the 10,000 people who would get their electricity from it. It might speed deployment for Hot Rock Energy to locate wells elsewhere and use the spare heat for co-located greenhouses and such.

255   For a survey of Australia's hot rock projects, see *www.pir.sa.gov.au/byteserve/petrol/prospectivity/apia_29_march07.pdf*

255   Photo credits: *Ormat.com* for Lyete, The Philippines. For the much-altered diagram, the MIT geothermal report of 2006.

257   Jon Gertner, "Atomic balm?" *New York Times* Magazine (16 July

2006).

259 Updated death toll for energy sources can be found at *www.uic.com.au/nip14app.htm*.

263 Advanced fast reactors: "If developed sensibly, nuclear power could be truly sustainable and essentially inexhaustible and could operate without contributing to climate change. In particular, a relatively new form of nuclear technology could overcome the principal drawbacks of current methods—namely, worries about reactor accidents, the potential for diversion of nuclear fuel into highly destructive weapons, the management of dangerous, long-lived radioactive waste, and the depletion of global reserves of economically available uranium." William H. Hannum, Gerald E. Marsh, George S. Stanford, "Smarter use of nuclear waste," *Scientific American* (December 2005): 84. At *gemarsh.com/wp-content/uploads/SciAm-Dec05.pdf*.

265 AC vs DC transmissions lines, illustration adapted from "Bulk power transmission at extra high voltages, a comparison between transmission lines for HVDC at voltages above 600 kV DC and 800 kV AC," an ABB Power Technologies presentation by Lars Weimers, n.d.

268 One interesting use of biofuels would be if they were burned for electricity and the CO_2 captured and sunk. It still has the hazards of CO_2 storage burps, and I cannot imaging it having the sheer capacity for sinking the accumulated atmospheric CO_2 that plankton enhancement would have. Biofuels news story by Stephen Leahy at *www.ipsnews.net/news.asp?idnews=38384*.

Chapter 20.   **Arming for a Great War**

272 Flannery quote at *www.news.com.au/heraldsun/story/0,21985,21225432-661,00.html*.

277 FDR in 1940: "I know that private business cannot be expected to make all of the capital investments required for expansion of plants and factories and personnel which this program calls for at once… [The] Government of the United States stands ready to advance the necessary money to help provide for the enlargement of factories, of necessary workers, the development of new sources of supply for the hundreds of raw materials required, the development of quick mass transportation of supplies."

A new "cost plus fixed fee" contract allowed the government to defray all costs essential to the execution of defense contracts and guarantee the contractor a profit through a fixed fee determined in advance. In other words, the government assumed primary financial responsibility for the mobilization process. From Doris Kearns Goodwin's *No Ordinary Time* (Simon and Shuster, 1994), 59.

Chapter 21.   **Get It Right on the First Try**

278   David Attenborough (2006), quoted at *books.guardian.co.uk/review/politicsphilosophyandsociety/0,,1945625,00.html*

280   Michael R. Raupach, Gregg Marland, Philippe Ciais, Corinne Le Quéré, Josep G. Canadell, Gernot Klepper, and Christopher B. Field. "Global and regional drivers of accelerating CO_2 emissions" *PNAS* 104 (10 June 2007): 10288–10293 at *dx.doi.org/10.1073/pnas.0700609104*.

288   Wikipedia Projects illustrate how a Recovery Manual could be done without top-down organization: *en.wikipedia.org/wiki/Wikipedia:WikiProject*.

288   Behavioral economics. A useful summary is by Teresa Tritch, "Helping people help themselves," *New York Times* (14 February 2007) at *select.nytimes.com/2007/02/14/opinion/15tlkingpoints.html*.

292   The quote is attributed to Edmund Burke.

292   Martin Luther King, Jr., "The Casualties of the War in Vietnam." Speech on 25 February 1967 in Los Angeles, California. See *www.stanford.edu/group/King/publications/speeches/unpub/670225-001_The_Casualties_of_the_War_in_Vietnam.htm*

About the Author

WILLIAM H. CALVIN

Born in 1939 in Kansas City, I grew up in real Middle America, though I now have an overlay from living in Seattle since 1962. I did a lot of journalism and photography before college, majored in physics at Northwestern University, then branched out into neurophysiology via studies at MIT, Harvard Medical School, and the University of Washington (Ph.D., Physiology & Biophysics, 1966). That biophysics background, plus a quarter-century of following the literature, is why I can talk shop with the climate scientists and oceanographers.

I'm now Affiliate Professor Emeritus at the University of Washington School of Medicine. I've had a long association with academic neurosurgeons and psychiatrists without ever having had to treat a patient. Most of my research has been about brain cells and circuits, along with the big-brain evolutionary history. I started paying attention to climate when trying to understand how our big brains evolved so rapidly during the Ice Ages. I've written fourteen books in twenty-eight years and have begun incorporating my photographs (many of which can be found via my website, *WilliamCalvin.org*).

Index